ENVIRONMENTAL IMPACT ANALYSIS

ENVIRONMENTAL IMPACT ANALYSIS

A New Dimension in Decision Making

R. K. Jain, Ph. D.   L. V. Urban, Ph. D.

G. S. Stacey, Ph. D.

Van Nostrand Reinhold Environmental Engineering Series

 VAN NOSTRAND REINHOLD COMPANY

NEW YORK   CINCINNATI   ATLANTA   DALLAS   SAN FRANCISCO

LONDON   TORONTO   MELBOURNE

Van Nostrand Reinhold Company Regional Offices:
New York   Cincinnati   Atlanta   Dallas   San Francisco

Van Nostrand Reinhold Company International Offices:
London   Toronto   Melbourne

Copyright © 1977 by Litton Educational Publishing, Inc.

Library of Congress Catalog Card Number: 77-443
ISBN: 0-442-28807-7

Manufactured in the United States of America

Published by Van Nostrand Reinhold Company
450 West 33rd Street, New York, N.Y. 10001

Published simultaneously in Canada by Van Nostrand Reinhold Ltd.

15 14 13 12 11 10 9 8 7 6 5 4 3 2 1

Library of Congress Cataloging in Publication Data

Jain, Ravinder Kumar, 1935–
    Environmental impact analysis

    (Van Nostrand Reinhold environmental engineering
series)
    1. Environmental impact analysis.   2. Environ-
mental impact statements.   I. Urban, Lloyd V.,
joint author.   II. Stacey, Gary S., 1940–
joint author.   III. Title.
TD170.2.J3   301.3   77-443
ISBN 0-442-28807-7

Van Nostrand Reinhold Environmental Engineering Series

THE VAN NOSTRAND REINHOLD ENVIRONMENTAL ENGINEER-
ING SERIES is dedicated to the presentation of current and vital information
relative to the engineering aspects of controlling man's physical environment.
Systems and subsystems available to exercise control of both the indoor and
outdoor environment continue to become more sophisticated and to involve
a number of engineering disciplines. The aim of the series is to provide books
which, though often concerned with the life cycle—design, installation, and
operation and maintenance—of a specific system or subsystem, are comple-
mentary when viewed in their relationship to the total environment.

Books in the Van Nostrand Reinhold Environmental Engineering Series
include ones concerned with the engineering of mechanical systems designed
(1) to control the environment within structures, including those in which
manufacturing processes are carried out, (2) to control the exterior environment through control of waste products expelled by inhabitants of structures and from manufacturing processes. The series will include books on
heating, air conditioning and ventilation, control of air and water pollution,
control of the acoustic environment, sanitary engineering and waste disposal,
illumination, and piping systems for transporting media of all kinds.

Van Nostrand Reinhold Environmental Engineering Series

ADVANCED WASTEWATER TREATMENT, by Russell L. Culp and Gordon L. Culp

ARCHITECTURAL INTERIOR SYSTEMS—Lighting, Air Conditioning, Acoustics, John E. Flynn and Arthur W. Segil

SOLID WASTE MANAGEMENT, by D. Joseph Hagerty, Joseph L. Pavoni and John E. Heer, Jr.

THERMAL INSULATION, by John F. Malloy

AIR POLLUTION AND INDUSTRY, edited by Richard D. Ross

INDUSTRIAL WASTE DISPOSAL, edited by Richard D. Ross

MICROBIAL CONTAMINATION CONTROL FACILITIES, by Robert S. Runkle and G. Briggs Phillips

SOUND, NOISE, AND VIBRATION CONTROL, by Lyle F. Yerges

NEW CONCEPTS IN WATER PURIFICATION, by Gordon L. Culp and Russell L. Culp

HANDBOOK OF SOLID WASTE DISPOSAL: MATERIALS AND ENERGY RECOVERY, by Joseph L. Pavoni, John E. Heer, Jr., and D. Joseph Hagerty

ENVIRONMENTAL ASSESSMENTS AND STATEMENTS, John E. Heer, Jr. and D. Joseph Hagerty

ENVIRONMENTAL IMPACT ANALYSIS: A NEW DIMENSION IN DECISION MAKING, R. K. Jain, L. V. Urban and G. S. Stacey

Preface

On January 1, 1970, the President of the United States signed the National Environmental Policy Act (NEPA) into law—a law encouraging productive and enjoyable harmony between man and his environment. The Act has brought about a profound impact on actions affecting the environment by requiring environmental considerations to be included in decision making processes concerning federal projects and activities. Specifically, NEPA requires each federal agency to prepare a detailed statement of environmental impact before proceeding with any major action, recommendation, or report on proposals for legislation that may significantly affect the quality of the human environment. Major actions may range from dredging operations to new highway construction, or from recommendation on legislation to major water resources projects. With the subsequent passage, by many state governments, of analogous legislation, environmental impact assessment has become an increasingly important vehicle for curbing pollution and minimizing environmental disruption.

But the environment is unbelievably complex; it encompasses not only the more obvious areas of air and water quality and ecology (which are highly complex), but less tangible areas as well, such as sociology, economics, land use, and aesthetics. One who attempts to analyze environmental impact on a solely individual basis soon finds himself at a loss. No individual possesses the skills and the range and depth of knowledge necessary to adequately address all aspects of the environment in an environmental impact analysis. Recognizing this fact, the authors of NEPA built into the law the requirement that the environmental studies be conducted in a systematic and interdisciplinary manner.

Just as no one individual can be proficient in all areas of the environment, no single text, including this one, can provide all the information required to completely encompass all areas. It can, however, provide readers with insight into each complex area of the environment and furnish guidance in the systematic and interdisciplinary approach to environmental impact analysis.

By emphasizing the interdisciplinary approach, the authors have attempted to produce a book useful to a wide range of readers—from biological and physical scientists to sociologists and economists; from students of engineering and environmental sciences to members of concerned citizens' participation groups; from consultants in the field to personnel at federal, state, and regional agencies involved in comprehensive planning and environmental impact analysis—anyone with a desire to achieve an understanding of man's impact on his environment.

The text covers a wide range of concepts involved in environmental impact analysis. Chapter 1 introduces the nature of the problem and lays the groundwork for the text's overall concept. Chapter 2 describes the different elements of NEPA, gives illustrative examples of executive orders and guidelines, and provides a commentary on state environmental policy acts. Chapter 3 discusses the differences between environmental impact assessments and statements (EIA/EIS), the legislative basis for preparing EIA/EIS, the function and purpose of the impact assessment process, and the types of actions covered by NEPA requirements. Also included is the requisite detailed content of EIA/EIS and general guidelines for processing these documents. Chapter 4 examines the four basic elements of environmental impact analysis, and Chapter 5 provides a review of impact assessment methodologies, as well as information regarding six basic categories of assessment methodologies. Chapter 6 provides a framework of a generalized approach for developing an overall impact analysis system on an agency-wide basis. This chapter also provides a rationale for a computer-based analysis system, a requirement necessitated by the complexity of the problem and the vast amount of information which must be processed for a meaningful environmental impact analysis. Since persons of various organizations and management levels review EIA/EIS documents, Chapter 7 presents a general procedure for reviewing EIA/EIS pertaining to construction-related projects; this general approach is applicable to other actions affecting the environment. There are numerous special issues which one can discuss in the overall context of environmental impact analysis; three of these, discussed in Chapter 8, deal with public participation, economic impact analysis, and energy impact analysis.

Three appendices are also included: Appendix A reproduces the National Environmental Policy Act; Appendix B provides a description of environmental attributes that are valuable for providing information regarding broad technical specialties and significant environmental attributes within these specialties. A comprehensive write-up, useful for environmental impact analysis, is provided for each attribute. Appendix C provides an illustrative example of a matrix-based, step-by-step procedure for preparing EIA/EIS.

Concepts discussed here have undoubtedly occurred to others and, in fact, may have been first expounded by others. Appropriate credits to the work of authors cited in this text are referenced. This text is a synthesis of many ideas dealing with environmental impact analysis and the new dimension in decision making they have created. Hopefully, this text will provide a framework and a thinking process for environmental impact analysis, as well as a meaningful tool for responding to NEPA and other related regulatory requirements, and for considering the environment as an integral part of project implementation.

Special credit must go to Dr. Maurice L. Warner, Dr. Neil L. Drobny, and Mr. Dale Manty for their work in Chapters 5, 7, and 8, respectively. We are indebted to them for their valuable inputs in these areas.

The support and encouragement provided by many of our colleagues at the U.S. Army Construction Engineering Research Laboratory (CERL), Champaign, Illinois; the University of Illinois, Urbana-Champaign; Texas Tech University, Lubbock, Texas; and Battelle-Columbus Laboratories, Columbus, Ohio are gratefully acknowledged.

Contents

ENVIRONMENTAL IMPACT ANALYSIS

I

Introduction to Environmental Impact Analysis

For many years, environmental considerations were ignored in the development of the United States. More recently, environmental factors have played a significant role in the speed and direction of our national progress. These factors are developing in our values a new concern and recognition of the dependence that we, as human beings, have on the long-term viability of the environment for sustaining human life. This new ethic of conservation of exhaustible resources has grown as concern for the environment has grown, because much of our environmental quality is itself a nonrenewable resource.

Man, in the twentieth century, represents an intrusion on the overall balance of processes that maintain the earth as a habitable place in the universe. We are recognizing this fact in our concern for the environment, but we are reluctant to give up our profligacy in the consumption of resources. Thus, it is incumbent upon mankind to examine his actions and to attune them so that they ensure long-term viability of earth as a habitable planet. Environmental impact assessments are a logical first step in this process, because they represent the opportunity for man to consider, in his decision making, the effects of actions that are not accounted for in the normal market exchange of goods and services. These effects need to be parlayed against the economic advantages derived from a given action. Adherence to pure economic exchange theory and practice for decision making has possible long-term adverse consequences for the planet Earth. Economic rules for decision making were adequate as long as the effects of man's activities were insignificant to the long-term viability of the planet as a place to reside. This book has been developed to instruct the user in how to consider the environmental consequences of his actions in compliance with environmental laws, regulations, and concerns. In addition, the spirit of compliance, namely a fuller consideration of noneconomic consequences of actions, is imparted to the user. A shopworn analogy would characterize the swing toward concern for environmental considerations as a pendulum that is on the verge of swinging back toward economic issue dominance. This type of trade-off is essential and is one that will always be made, but mankind must be aware that sacrificing long-term viability for short-term expediency is less than a bad solution; it is no solution.

1.1 WHAT IS ENVIRONMENTAL IMPACT ANALYSIS?

In order to incorporate environmental amenities into a decision or a decision making process, it is necessary to develop a complete understanding of the possible and probable consequences of a proposed action on the environment. However, prior to this development, a clear definition of the environment must be constructed.

The word "environment" means many different things to different people. To some, the word conjures up thoughts of woodland scenes with fresh, clean air and pristine waters. To others, it means their man-modified neighborhoods or immediate surroundings. Still others relate "environment" to "ecology," and think of plant-animal interrelationships, food chains, threatened species, and so forth.

Actually, the "environment" is a combination of all these concepts plus many, many more. It includes not only the areas of air, water, plants, and animals, but also other natural and man-modified features which constitute the totality of our surroundings. Thus, transportation systems, land use characteristics, community structure, and economic stability all have one thing in common with carbon monoxide levels, dissolved solids in water, and natural land vegetation—they are all characteristics of the environment. In other words, the environment is made up of both biophysical *and* socioeconomic elements which sould be considered in environmental impact analysis.

But what is meant by "impact" analysis? Simply stated, "impact" means change—any change, positive or negative—from a desirability standpoint. An environmental impact analysis is, therefore, a study of the probable changes in the various socioeconomic and biophysical characteristics of the environment which may result from a proposed or impending action.

In order to accomplish the analysis, it is first necessary to develop a complete understanding of the proposed action. What is to be done? What kinds of materials, manpower, and/or resources are involved?

Second, it is necessary to gain a complete understanding of the affected environment. What is the nature of the biophysical and/or socioeconomic characteristics that may be changed by the action?

Third, it is necessary to project the proposed action into the future and to determine the possible impacts on the environmental characteristics, quantifying the changes whenever possible.

Fourth, it is necessary to report the results of the analysis in a manner such that the analysis of probable environmental consequences of the proposed action may be used in the decision making process.

The exact procedures to be followed in the accomplishment of each environmental impact analysis are by no means simple or straightforward. This is due primarily to the fact that many and varied projects are proposed for equally numerous and varied environmental settings. Each combination results in a unique cause-condition-effect relationship, and each combination must be studied individually in order to accomplish a comprehensive analysis. Generalized procedures for conducting an analysis in the manner indicated by the four steps indicated above have been developed. These procedures will be explained in subsequent chapters of this book.

1.2 WHO ACCOMPLISHES ENVIRONMENTAL IMPACT ANALYSIS?

The necessity for preparing an environmental impact analysis may vary with individual projects or proposed actions. For many actions, there is a legal basis for conducting such an analysis. For others, the environmental analysis may be undertaken simply for incorporation of environmental considerations into planning and design, recognizing the merit of such amenities on an economic, aesthetic, or otherwise desirable basis.

In the first five years after the enactment of the National Environmental Policy Act (NEPA) on January 1, 1970, some 6000 environmental impact statements have been filed for federal agency actions.[1] The Department of Transportation consistently has been the single agency filing the largest number of statements followed by the Corps of Engineers, the Department of Agriculture, and the Atomic Energy Commission (see Fig. 1.1). Thus, it may be seen that the kinds of projects most frequently undergoing environmental analysis include roads and highways, watershed and flood control projects, airports, energy-related projects, and parks and wildlife refuges.

The individuals who actually prepare the statements for the agencies represent a wide variety of disciplines and expertise, and address the many components of the environment. Working as a team, biologists, economists, engineers, sociologists, planners, architects, and other specialists (or generalists) approach environmental impact analysis on an interdisciplinary basis. A typical study may involve the collection of data from both the field and existing sources. On the basis of these data and the judgment of the analysts, projections of changes in the environment are made, and the documentation of the projections are finalized. Specific procedures for this process have been developed and will be discussed in detail in subsequent chapters of this book.

1.3 PROCEDURES FOR ACCOMPLISHING ENVIRONMENTAL IMPACT ANALYSIS

This book has been written in response to a perceived need to consolidate the current state-of-the-art in environmental impact analysis in a concise document. Clearly, impact assessments themselves have, in many cases, exceeded the length of this book by orders of magnitude. There are undoubtedly approaches to assessment that are not explicitly discussed in this text. Nevertheless, this book represents a pathway through the impact analysis morass that engulfs the person faced with the obligation of preparing a statement reflecting the effects of a federal agency's projected activities on the environment.

Because the book is in a monograph form, it does not relieve the reader of the responsibility or necessity to think about his problems. As a matter of fact, the purpose of NEPA is to stimulate the respondents to be conscious and concerned about the environment. Therefore, the reader will not find quick answers to all questions in this text, but instead will learn a *process* for considering what the effect of human actions will be on the environment.

The objective of the material, as presented in this text, is to show the reader how an analysis of a project's effect on the environment can be developed.

[1]*Environmental Quality*, Fifth Annual Report of the Council on Environmental Quality, p. 389, U.S. Govt. Printing Office, December 1974.

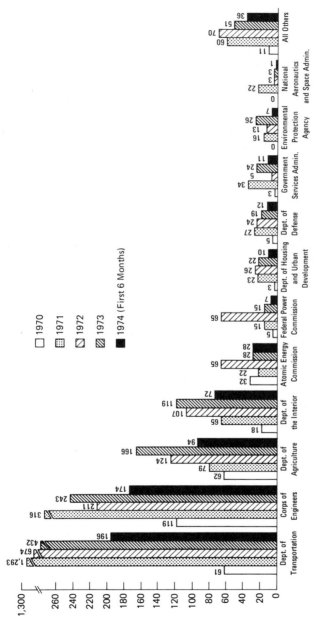

Fig. 1.1 Environmental impact statements filed annually, by agency (CEQ 5th Annual Report).

There are several initial difficulties. First, one must decide where and how to begin. Second, one needs to know what steps to take (from an analysis perspective) and the results that are being sought. As the analysis proceeds, questions will arise regarding whether all relevant subjects have been examined. This text provides the framework within which the analyst can be systematic about examining environmental impacts and can feel some confidence, upon completion, that all major possible effects of a project have been considered.

When an analysis of a project's environmental effects is required, one is faced with the task of deciding how much resources (or money) must be devoted to performing the analysis at the required level. This decision can be made more easily if it is known how to perform the analysis and if the likelihood of the analysis revealing adverse impacts, which may reduce the probability of having the project funded, is considered. While this text will not answer the question of how much an analysis will cost, it will indicate what must be done, and the reader can use this information to guide himself in estimating the cost of preparing an analysis.

In many cases, environmental impact assessments have been performed with strict adherence to rules and regulations set forth by the subject agency. Unfortunately, in the process of adhering to the rules in a mechanical manner, the spirit of NEPA is sometimes missed, and the assessment may depart from a functional exercise to a nonfunctional paper exercise having no substantial meaning. In these cases, the analysis is worse than useless—it is actually of disutility to the preparer, the impacted group, and the receiving agencies. Therefore, instead of being a "cookbook," this text presents an approach that directly relates the results of being responsive to both the letter and the spirit of the NEPA and the President's Council on Environmental Quality (CEQ) guidelines.

2

National Environmental Policy Act

On January 1, 1970, the President of the United States signed the National Environmental Policy Act (NEPA), PL 91-190, into law. The enactment of this legislation established a national policy of encouraging productive and enjoyable harmony between man and his environment. The symbolism of the timing of this law did not go unnoted by the President and other concerned Americans, who heralded the 1970's as a decade of environmental concern. Enactment of NEPA and concern regarding the environment and quality of life among people around the world have generated significant environmental protection legislation and regulations in many industrialized nations besides the United States. Provisions and policies set forth in NEPA are being emulated by many states within the United States and within other nations as well.[9]

The main purposes of this legislation, as set forth in the Act, are:[7] "To declare a national policy which will encourage productive and enjoyable harmony between man and his environment; to promote efforts which will prevent or eliminate damage to the environment and biosphere and stimulate the health and welfare of man; to enrich the understanding of the ecological systems and natural resources important to the Nation; and to establish a Council on Environmental Quality."

2.1 ELEMENTS OF NEPA

There are two titles under this Act[7]: Title I, Declaration of National Environmental Policy, and Title II, Council on Environmental Quality (CEQ).

Title I sets forth the national policy on restoration and protection of environmental quality. The relevant sections under this title are summarized as follows:

Section 101.   Requirements of Section 101 are of a substantive nature. Under this section, the federal government has a continuing responsibility "consistent with other essential considerations of national policy..."[7] to minimize adverse environmental impact and to preserve and enhance the environment as a result of implementing federal plans and programs.

Section 102.   Section 102 requirements are of a procedural nature. Under this section, the proponent federal agency is required to make a full and adequate analysis of all environmental effects of implementing its programs or actions.

In Section 102(1), Congress directs that policies, regulations, and public laws shall be interpreted and administered in accordance with the policies of NEPA; Section 102(2) directs all federal agencies to follow a series of steps to ensure that the goals of the Act will be met.

The first requirement is found in Section 102(2)(A), where it is required that "a systematic and interdisciplinary approach..."[7] be used to ensure the integrated use of social, natural, and environmental sciences in planning and decision making.

Section 102(2)(B) states that federal agencies shall, in consultation with CEQ, identify and develop procedures and methods such that "presently unquantified environmental amenities and values may be given appropriate consideration in decision making..."[7] along with traditional economic and technical considerations.

Section 102(2)(C) sets forth the requirements and guidelines for preparing environmental impact statements (EIS). This section requires all federal agencies to include in every recommendation or report on legislative proposals and other major federal actions significantly affecting the quality of the human environment, a detailed statement by the responsible official covering the following elements:[7]

1. The environmental impact of the proposed actions;
2. Any adverse environmental effects which cannot be avoided should the proposal be implemented;
3. The alternatives to the proposed action;
4. The relationship between local short-term uses of man's environment and the maintenance and enhancement of long-term productivity; and
5. Any irreversible and irretrievable commitments of resources which would be involved in the proposed action should it be implemented.

Specific EIS format, coordination, instruction, approval, and review hierarchy are established by each federal agency. Persons preparing an EIS should follow the instructions of their organizations.

Since the enactment of this Act, CEQ has issued various guidelines to clarify the requisite content of an environmental impact statement. The latest guidelines for the preparation of environmental impact statements were issued by CEQ in August 1973.[8] These guidelines clarify the content of environmental impact statements, concerning types of impacts to be covered, points to be discussed, appropriate alternatives to be evaluated, the extent of secondary impacts to be included, and requirements for negative declaration decisions. The eight major points to be covered by impact statements, as outlined in the latest CEQ guidelines, are summarized as:[8]

1. A description of the proposed action, a statement of its purpose, and a description of the environmental setting of the project;
2. The relationship of the proposed action to land use plans, policies, and controls for the affected area;
3. The probable impact of the proposed action on the environment;
4. Alternatives to the proposed action, *including* those not within the existing authority of the responsible agency;
5. Any probable adverse environmental effects that cannot be avoided (summarizing the unavoidable impacts [point 3], and stating how each avoidable impact [point 3] will be mitigated);
6. The relationship between local short-term uses of man's environment and the maintenance and enhancement of long-term productivity;
7. Any irreversible and irretrievable commitments of resources (including natural and cultural as well as labor and materials);
8. An indication of what other interests and considerations of federal policy are thought to offset the adverse environmental effects identified.

Chapter 3 provides a detailed discussion of these eight points.

Section 103.   This section requires all federal agencies to review their regulations and procedures, "for the purpose of determining whether there are any deficiencies or inconsistencies therein which prohibit full compliance with the purposes and provisions of this Act and shall propose to the President. . .such measures as may be necessary to bring their authority and policies into conformity with. . .this Act."[7]

Title II establishes the Council on Environmental Quality (CEQ) as an environmental advisory body for the executive office. In addition, the President is required to submit to the Congress an annual "Environmental Quality Report." This yearly summary sets forth (1) the status and condition of the major natural, man-made, or altered environmental classes of the nation; (2) current and foreseeable trends in the quality, management, and utilization of such environments and socioeconomic impacts of these trends; (3) the adequacy of available natural resources; (4) a review of governmental and nongovernmental activities on the environment and natural resources; and (5) a program for remedying the deficiencies and recommending appropriate legislation.[7]

2.2   EFFECTS OF NEPA

Effects of NEPA have been far-reaching. This Act, in many instances, has been instrumental in requiring reassessment of many federal programs (and programs where federal participation, approval, or license is involved)—both newly proposed programs and ongoing programs in various stages of completion and implementation. In the reassessment process, federal agencies have been required to consider not only the economic and mission requirements, but also both the positive and negative environmental impacts.

When the environmental costs, as surfaced because of the requirements of NEPA (i.e., documentation of an EIS), are made known to the decision makers at various official levels and to the public, modification, delay, or abandonment of the project may be necessitated. In some cases, the modification or abandonment of the project will be made at the federal agency's own initiative; however, in most cases, strong pressure from the public, environmental groups, and court actions will be the driving forces.

Judicial Review.   Initially, the court cases resulting from NEPA dealt primarily with procedural requirements of the Act. Most of these basic procedural questions have now been settled. Consequently, litigation in 1972 and 1973 had turned to the content of statements and, more recently, to the agency decisions made after statements are completed.

In one case, (*Sierra Club versus Froehlke*, [6] February 1973), a Federal District Court enjoined the U. S. Army Corps of Engineers from proceeding with the Wallisville Dam Project becuase of the inadequacy of the EIS content. The court concluded that (1) the statement did not adequately disclose its relationship to the much larger project (Trinity River Project); (2) the statement lacked the requisite detail to satisfy the Act's full disclosure requirement; (3) alternatives to the project were inadequately considered; and (4) there was no indication that genuine efforts had been made to mitigate any of the major impacts on the environment.

Another court case (*Natural Resources Defense Council versus Grant*[6]) involved a watershed project by the Soil Conservation Service in North Carolina, USA. The court held that the impact statement omitted or inadequately described many of the important environmental effects of the project and failed to discuss adequately alternatives to the project. The court's conclusion was that NEPA's full disclosure requirements were not met by this impact statement.

In addition, several recent court cases have confirmed the role of the judicial branch of the U. S. government in reviewing the substance of the agency decisions. Affirmation of this judicial role came in the Gillham Dam case, in which the Court of Appeals concluded that there is a judicial responsibility to make sure that an agency has not acted "arbitrarily and capriciously" in making decisions affected by NEPA.[6]

While many of the remaining controversial issues dealing with the interpretation and implementation of NEPA will have to be settled in courts in the coming years, it can be stated at this time that the court decisions have established precedents beyond the procedural requirements of preparing an environmental impact statement as required by NEPA.

The proponent agencies, as a result of court cases and judicial review, and in order to comply with the requirements of NEPA, must also do the following:

1. Satisfy the Act's full disclosure requirements with adequate detail (i.e., include significant environmental impacts and the relationship of the project assessed to other related projects);
2. Adequately consider the alternatives to the project; and
3. Make genuine efforts to mitigate any major impacts on the environment due to implementation of the project.

As a result of recent court cases and judicial review, it appears that the courts have a judicial responsibility to review the substance of the statements and the agency decision, based on the statement, to make sure that the agency has not acted arbitrarily and capriciously in making decisions when dealing with environmental considerations.

Beneficial Effects.   It appears that there will always be some controversy concerning the beneficial effects of NEPA on the environment and the quality of life and the economic costs associated with the implementation of this Act. Regardless of the controversy, NEPA, in setting forth national policy on restoration and protection of environmental quality, has declared that it is a continuing policy of this government, in cooperation with state and local governments and other concerned public and private organizations, to create and maintain conditions under which man and nature can exist in productive harmony and fulfill social, economic, and other requirements of present and future generations of Americans.

As summarized by a CEQ annual report,[5] the beneficial effects of NEPA are:

1. To bring national policies in line with modern concerns for environmental protection and enhancement of environmental quality;
2. To provide a systematic way of dealing with problems that transcend the parochial interests of individual federal agencies and various interested groups;
3. To open governmental activities to public scrutiny and public participation by fully disclosing the environmental costs associated with the federal action;
4. To staff governmental agencies with personnel capable of undertaking the interdisciplinary approach, and to analyze the environmental costs involved as required by NEPA;
5. To allow for citizen suits to provide for the enforcement of requirements of NEPA; and
6. To provide a vehicle to include environmental costs in the decision making process.

2.3   IMPLEMENTATION OF NEPA

It must be noted that NEPA and its implementation have not been without their critics (as perhaps typified by Paul Ehrlich's article, entitled "Dodging the Crisis").[4] Considerable litigation has developed concerning compliance with (or, in the view of some, circumvention of) the provisions of the Act. Notable among these was the "Calvert Cliffs Case" in which the courts held that compliance with established environmental standards did not relieve a governmental agency from the NEPA requirement of considering all environmental factors when assessing impact. In this case, the Atomic Energy Commission had sought to exclude water quality considerations from its assessment of the impact of a nuclear power plant, on the grounds that a state had certified compliance with water quality standards under the relevant federal water pollution control legislation.

Among the frequently voiced concerns about the implementation of NEPA[2] are (1) that impact statements are not available in time to accompany proposals through review procedures; (2) that statements are prepared in "mechanical compliance" with NEPA; (3) that impact statements are biased to meet the needs of predetermined program plans; (4) that agencies may disregard the conclusions of adverse impact statements; (5) that CEQ lacks authority to enforce the intent of NEPA; (6) that intangible environmental amenities are being ignored; (7) that secondary effects are being ignored; and (8) that inadequate opportunity is available for public participation and reaction.

Perhaps the most severe of these reservations concerning NEPA was summarized by Roger C. Crampton, who testified that "the agencies must guard against a natural but unfortunate tendency to let the writing of impact statements become a form of bureaucratic gamesmanship, in which the newly acquired expertise is devoted not so much to shaping the project to meet the needs of the environment, as to the shaping of the impact statement to meet the needs of the agency's preconceived program and the threat of judicial review."[3] Perhaps the point which should be made is that an impact statement for a project should not be used as a justification for a preconceived program, but rather it should be used as a vehicle for a full disclosure of the potential environmental impacts involved. Also, it should be used as a tool for adequately considering the environmental amenities in decision making and for allowing participation in the project by other federal and state agencies and the public, to provide proper consideration of the environment, along with economic and project objective requirements.

2.4 COUNCIL ON ENVIRONMENTAL QUALITY (CEQ)

Title II of NEPA created, in the Executive Office of the President of the United States, a Council on Environmental Quality (CEQ). This Council is composed of three members, who are appointed by the President with the advice and consent of the Senate. The President designates one of the members of the Council to serve as chairman. In addition, the Council employs environmental lawyers, professional scientists and other employees to carry out its functions as required under NEPA. Duties and functions of CEQ may be summarized as follows:[7]

1. Assist and advise the President in the preparation of the Environmental Quality Report as required by NEPA;
2. Gather, analyze, and interpret, on a timely basis, information concerning the conditions and trends in the quality of the environment, both current and prospective;
3. Review and appraise the various programs and activities of the federal government in light of the policy of environmental protection and enhancement, as set forth under Title I of NEPA;

4. Develop and recommend to the President national policies to foster and promote improvement of environmental quality to meet many goals of the nation;
5. Conduct research and investigations related to ecological systems and environmental quality;
6. Accumulate necessary data and other information for a continuing analysis of changes in the national environment and interpretation of the underlying causes;
7. Report at least once a year to the President on the state and condition of the environment; and
8. Conduct such studies and furnish such reports and recommendations as the President may request.

A significant feature to note is that both the charter assigned and the responsibilities delegated to CEQ are quite extensive. However, CEQ does not have any regulatory or policing responsibilities, but is highly influential in its advisory capacity.

2.5 EXECUTIVE ORDERS AND AGENCY RESPONSE

To further enhance and explain NEPA, several Executive Orders were issued by the President, and the federal agencies responded with appropriate guidelines and directives. As an illustration, excerpts from some of the executive orders and agency responses are presented herein.

Executive Order 11752, "Prevention, Control, and Abatement of Air and Water Pollution at Federal Facilities," 17 December 1973. Section 1. Policy. It is the purpose of this order to assure that the federal government in the design, construction, management, and maintenance of its facilities shall provide leadership in the nationwide effort to protect and enhance the quality of our air, water, and land resources. . . .

Executive Order 11602, 30 June 1971. Sets the policy with respect to federal contracts, grants, or loans for the procurement of goods, materials, or services as being undertaken in such a manner that will result in effective enforcement of the Clean Air Act Amendments of 1970.

Executive Order 11514, "Protection and Enhancement of Environmental Quality", 5 March 1970. Section 1. Policy. The federal government shall provide the leadership in protecting and enhancing the quality of the nation's environment to sustain and enrich human life. Federal agencies shall initiate measures needed to direct their policies, plans and programs so as to meet national environmental goals. The Council on Environmental Quality, through the chairman, shall advise and assist the President in leading this national effort. . . .

Also, the heads of federal agencies are required to monitor, evaluate, and control, on a continuing basis, their agencies' activities so as to protect and enhance the quality of the environment. . . .

Agency Responses

Nearly all federal agencies have issued directives, guidelines, circulars and other appropriate documents in response to Executive Orders and NEPA. Because of the changing nature of these documents, it would be infeasible to include extensive information about them. A summary of some of these guidelines has been prepared by Dickert and Domeny.[10] It is suggested that the current appropriate agency information be consulted prior to embarking upon an environmental impact analysis.

2.6 STATE ENVIRONMENTAL POLICY ACTS

Because of the concern for environmental protection and enhancement, NEPA was enacted by Congress and signed into law by the President in 1970. NEPA applied to the activities and programs of the federal agencies and to those activities and programs supported by federal funds. Many states felt that, in many instances, the problems and concerns at the state level were different from those at the federal level, and that they varied from one state to the other. Since many state-supported projects were not covered by the requirements of NEPA, some of the states enacted their own state environmental policy acts or guidelines, sometimes referred to as State Environmental Policy Acts ("SEPA's") or "little NEPA's."

Nearly half of the states had enacted state environmental policy acts by 1975.[6] Each year, additional states (and even some cities) join the ranks of those that have already enacted state environmental policy acts. Since state environmental policy acts are patterned after NEPA, discussion and procedures presented in this text can be used to address impact analysis requirements set forth by the states.

REFERENCES

1. *Calvert Cliff's Coordinating Committee versus Atomic Energy Commission,* 499 F. 2nd 1109, 2ERC1779, 1ELR2036, D.C. Cir 1971 (West, 1971).

2. *CF Letter*, The Conservation Foundation, May 1972.

3. Crampton, R. C., Testimony before Joint Meeting of the Interior and Public Works Committees of U. S. Senate, March 7, 1972.

4. Ehrlich, P., "Dodging the Crisis," *Saturday Review* **53**, *No. 11:* 73 (November 1970).

5. *Environmental Quality*, Third Annual Report of the Council on Environmental Quality, U. S. Government Printing Office (USGPO), August 1972.

6. *Environmental Quality*, Fourth Annual Report of the Council on Environmental Quality, U. S. Government Printing Office (USGPO), pp. 237–238, September 1973.

7. National Environmental Policy Act of 1969 (PL 91-190; 83 Stat 852).

8. Preparation of Environmental Impact Statements: Guidelines, *Federal Register* **38**, *No. 147*, Part II: 20550-20562 (August 1, 1973).

9. Munn, R. E., *Environmental Impact Assessment: Principles and Procedures*, United Nations, SCOPE, Secretariat, 51, Boulevard de Montmorency 75016 Paris, France. (Report 5, Toronto, Canada) 1975.

10. Dickert, G. and K. R. Domeny, *Environmental Impact Assessment: Guidelines and Commentary*, University Extension, University of California, Berkeley, 1974.

3

Environmental
Impact Assessments
and Statements

The environmental impact assessment (EIA) and the environmental impact statement (EIS) differ somewhat in purpose and use. This chapter discusses these differences and outlines procedures and steps for preparing and processing these documents. The discussion focuses mainly on NEPA: however, since all state and some major city environmental policy act requirements parallel those of NEPA, the information presented herein is also relevant to the state and local environmental policy acts.

3.1 DEFINITIONS AND DISCUSSION OF EIA/EIS

While it is the common goal of EIA and EIS documents and the process leading to their development, to ensure that environmental considerations are made part of the agency (federal, state, or local) decision making process, these documents are utilized for different purposes to achieve that end. The EIA may be defined as the documentation of an environmental analysis, which includes identification, interpretation, prediction, and mitigation of impacts caused by a proposed action or project. This document (EIA) is utilized to provide a basis for intra-agency review of project impacts, and is designed to provide adequate information for judging whether an environmental impact statement should be prepared.

An EIS may be defined as the documentation of an environmental analysis of a project or action with a potential for environmental impacts which are either significant or highly controversial. Under NEPA, federal agencies are required to prepare and use an EIS in their agency review and decision-making process and to submit the statement to CEQ. This is to be accomplished before the agency undertakes any major action that significantly affects the quality of the human environment or which is controversial for environmental reasons. The federal agency responsible for preparing an EIS is the proponent agency.

The EIS document should follow the format and content as provided by current CEQ guidelines and proponent agency requirements. The first step in the development of this document is the preparation of a "draft EIS." After inter-agency and public review, a final EIS is prepared, which addresses opposing responsible views expressed by other agencies and the public.

The broad spectrum of federal (or state or local) actions ranges from minor to major. However, most are not major (from the standpoint of constituting a significant commitment of resources), and may not have a significant environmental impact. Where it is apparent that the proposed action is minor, and that there will be no adverse environmental impact, a simple declaration of that fact is usually sufficient for record purposes. Such a declaration should be made early in the planning process and be made a part of the planning documentation on the proposed action. In other words, the planner should include a statement in his written records that an environmental assessment was made, and that it was concluded that the action is not major, nor will it have a significant impact on the environment.

Between those actions having an obvious and significant environmental impact requiring an EIS, and those having nearly none, there are a large variety of actions where it is difficult to determine, off-hand, the extent of their environmental impacts. To prevent a planner from resorting to snap judgments in such instances, an environmental impact assessment should be prepared. This is an analysis much like an EIS, but it is not prepared in the same depth, nor need it include all the elements of an EIS. For example, it may be unnecessary to pursue a rigorous examination of various alternative courses of action when the analysis of a preferred course of action reveals that there is no significant impact on the environment or that the action is not controversial. However, any assessment of an action should identify all direct and long-term adverse impacts and any irreversible commitments of resources (ecological and material) that may result from proceeding with the action. Figure 3.1 indicates the EIA/EIS relationship in the environmental analysis process.

A properly prepared assessment should enable the planner to conclude whether the proposal should or should not be regarded as a major action, whether the environmental impact is or is not significant, and if the action could be environmentally controversial. Whenever it is concluded that significant environmental impact will result from a proposed action, or that it may become environmentally controversial when others learn of the action, a draft EIS must be prepared.

As a word of caution, it is not prudent for a planner to avoid preparing an EIS by understating the possible impact of the action. An important project could be halted or seriously delayed by approving authorities at higher levels who may question the validity of a determination that there is no environmental impact. Similarly, doubts could be raised by others outside the agency, including private citizens, who may exercise their right

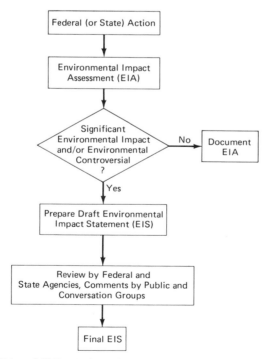

Fig. 3.1   EIA and EIS relationship in the environmental analysis process.

to legally enjoin the action until the environmental impact of the action is fully disclosed.

3.2 LEGISLATIVE BASIS

As discussed in Chapter 2, Section 102(2)(C) of NEPA forms the legislative basis of the requirement for preparing environmental impact statements for major federal actions. In essence, to ensure that environmental amenities and values are given systematic consideration equal to economic and technical considerations in the federal decision making process, NEPA requires that each federal agency prepare a statement of environmental impact before each major action, including, but not limited to, recommendations for legislation; new and continuing project and program activities; and the making, modification, or establishment of regulations, rules, procedures, and policies that may significantly affect the quality of the human environment.[3,4]

Initially, the requisite content of an EIS was outlined in Section 102(2)(C) of NEPA. Subsequent guidelines issued by CEQ have served to clarify EIS content requirement and have provided agencies with additional assistance in responding to the requirements set forth by this Act.

3.3 FUNCTION AND PURPOSE OF THE IMPACT ASSESSMENT PROCESS

The process of environmental impact analysis serves to meet the primary goal of Congress in enacting NEPA—to establish a national policy in favor of protecting and restoring the environment. The environmental impact statement process was included in NEPA to achieve a unified response from all federal agencies to the policy directives contained within the Act.

The primary purpose for preparing environmental impact statements is to disclose the environmental consequences of a proposed action, thereby alerting the agency decision maker, the public, and ultimately Congress and the President to the environmental risks involved. An important and intended consequence of this disclosure is to build into the agency's decision making process a continuing consciousness of environmental considerations.

Environmental impact assessment should be undertaken for reasons other than to simply conform to the procedural requirements of the law. According to the letter of the law, environmental impact must be assessed for activities with significant impact. However, the spirit of the law is founded on the premise that to utilize resources in an environmentally compatible way, and to protect and enhance the environment, it is necessary to know how activities will affect the environment, and to consider these effects early enough so that changes in plans can be made if the potential impacts warrant them.

In standard cost-benefit analysis and program evaluation, the intangible impacts on the environment are not taken into account. The impact assessment process provides the basis for operating within the spirit of the law by encouraging recognition of impacts early in the *planning* process and by providing an *inventory* of potential environmental effects of man's activities.

The planning process inevitably involves projecting activities into the future to determine how well the projected activities conform to anticipated alternative functions. The methods for dealing with short-term exigencies and complexities can be identified only with reference to the long-term plan.

Environmental impact assessment fits into the long-term planning process because it provides the vehicle for identifying the potential effects of activities on the environment. While immediate knowledge of these effects is important, the long-term aspects of impact are probably more important, because only on a longer time horizon can adequate, effective, and low-cost alternatives to reduce the impact be identified.

If, for example, the potential for an adverse impact of an activity or program planned for five years in the future were identified, adequate time to consider significant mitigation alternatives (including stopping the program) would exist. This is much preferred to finding out about serious impacts after an activity is half completed. In the latter case, modifications to reduce the impact could be very costly, or opposition could force costly delays in completion or even prevent continuation.

Historically, few records have been maintained of the long-term environmental effects of activities. EIA/EIS provide a vehicle for recording impacts of activities so that knowledge of what adverse changes may occur can be collected and maintained. The purpose of the inventory is to ensure disclosure of the impacts so that concerned institutions or individuals will be aware of possible repercussions of the subject activity.

Another valuable use for the inventory of impacts is to identify the potential cumulative effects of a group or series of activities in an area. Any single activity might not be likely to cause serious changes in the environment, but when its effects are added to those of other projects, the impacts on the environment might be severe. The potential for cumulative impacts must be identified, and in some cases, this may be possible only at the intra-agency level. Thus, to account for cumulative impacts, it might be more desirable to assess the environmental impact at a program level which covers many projects or activities.

Again, the National Environmental Policy Act has the primary goal of incorporating environmental considerations into the decision making process. NEPA cannot be used, nor was it intended to be used, to stop projects, provided the requirements of the Act are fulfilled. The essence of these requirements is simple: use a systematic and interdisciplinary approach to evaluate the environmental consequences of the proposed action, give appropriate consideration to unquantifiable environmental amenities, and incorporate the results into the decision making process. If this is done in a complete, honest, and straightforward manner, NEPA requirements are satisfied. The project or action may be an environmental disaster; however, if the probable consequences are known, fully disclosed, and weighed with other aspects of the action, then the letter and the spirit of NEPA have been fulfilled. It is only when the environmental assessment procedure is looked upon as a "paper exercise," or when the assessment is done in an incomplete or short-sighted manner, that legal difficulties can arise.

3.4 TYPES OF ACTIONS COVERED BY NEPA

As outlined in the August 1973 CEQ guidelines,[4] specific types of actions covered by NEPA are:

(a) Recommendations or favorable reports relating to legislation, including requests for appropriations. This includes agency recommendations for legislation concerning their own proposals and agency reports about legislation initiated elsewhere. Since only one agency is required to file an EIS, in the latter case, only the agency which has primary responsibility for the declared proposal will prepare an EIS. (b) New and continuing projects and program activities directly undertaken by federal agencies or supported, in whole or in part, through federal contracts, grants, subsidies, loans, or other forms of funding assistance. This also includes those projects which involve a federal lease, permit, license certificate, or other entitlement for use. Exceptions are those projects which are undertaken solely under the general revenue sharing funds distributed under the State and Local Assistance Act of 1972. Some states, however, may require preparation of an EIS under the provisions of the State Environmental Policy Acts. (c) The making, modifications, or establishment of regulations, rules, procedures, and policy.

3.5 DETAILED CONTENT OF EIA/EIS

It is well to reiterate that recent CEQ guidelines[4] emphasize that environmental considerations should be taken into account from the beginning of the decision making process. Initial environmental studies, for example, should be undertaken concurrently with initial technical and economic studies. Too often, in the past, assessments and statements have been written to justify decisions long since made. If environmental assessments had been initiated at the inception of the projects, environmental information could have been integrated into, rather than tacked onto, the decision making process and, in many cases, delays could have been avoided.[2] The guidelines require that draft impact statements (hence, the required prior assessments) are to be prepared and circulated at the earliest possible stage in the decision making process.

An important and intended consequence of preparing an EIA/EIS is to build into a federal agency's decision making process a continuing consciousness of environmental considerations. In order to do this, it must be emphasized again that *considerations of environmental factors must be integrated into existing agency procedures, and environmental factors must be considered from the very beginning of the planning process.* Therefore, an agency should prepare an EIA as early in the decision making process as possible, and in all cases, *prior* to an agency decision. *It is a misconception to think that an EIA is required only for construction projects.* As discussed previously, an EIA is required for all major federal actions which may affect the environment.

For preparing an EIS, it is mandatory that the five points of NEPA, Section 102(2)(C) (and further amplified by the eight points for EIS content set forth by CEQ guidelines*) be addressed and documented; however, for preparing an EIA, addressing these points is not mandatory. It is recommended, however, that for preparing an EIA, the eight points of CEQ be addressed in format similar to that of the EIS. The reason for this is simply that if an EIS would be required later, the procedure would be much further advanced if the assessment had conformed to the EIS format.

The subsequent paragraphs of this section contain specific information regarding the preparation and content of the EIA and the EIS. The following items are presented and discussed:

Sample cover sheet
Summary sheet
Outline for CEQ-prescribed EIS format
Detailed content of EIA/EIS (following the outline).

Figure 3.2 is a sample cover sheet, and Fig. 3.3 shows the format and content prescribed for the summary sheet. Figure 3.4 is the outline for CEQ-prescribed EIS content.

---

*The function of the CEQ guidelines is to interpret or translate the broad language of NEPA's Section 102 into more concrete terms that provide a degree of certainty without relieving agencies of the many critical judgments that they must make.

DEPARTMENT OF TRANSPORTATION
FEDERAL HIGHWAY ADMINISTRATION

Prepared by

(State) HIGHWAY DEPARTMENT

FINAL

ENVIRONMENTAL STATEMENT

ADMINISTRATIVE ACTION

For

STATE HIGHWAY (No.)

FROM (Location)
TO (Location)

(Name) COUNTIES

THIS HIGHWAY IMPROVEMENT IS PROPOSED FOR FUNDING UNDER TITLE 23, U.S.C.
THIS STATEMENT FOR THE IMPROVEMENT WAS DEVELOPED IN CONSULTATION WITH
THE FEDERAL HIGHWAY ADMINISTRATION AND IS SUBMITTED PURSUANT TO:

SECTION 102(2)(C)
PUBLIC LAW 91-190

_____   _____(Signature)_____   DEPARTMENT OF TRANSPORTATION
Date   District Engineer, (State) Hwy. Dept.

_____   _____(Signature)_____
Date   Chief Engineer of Highway Design,
              (State) Highway Department

REVIEWED FOR CONTENT AND ACCEPTED BY THE FHWA

_____   _____
Date   FEDERAL HIGHWAY ADMINISTRATION

Fig. 3.2   Sample cover sheet for Department of Transportation EIS.

The following information indicates details to be included in an EIS that is responsive to the CEQ-prescribed content. The discussion follows the outline presented in Fig. 3.4.

1. *Project Description.* Here, the action, activity, or project is described in sufficient detail to provide a reviewer unfamiliar with the proposed

```
┌─────────────────────────────────────────────────────────────────────────┐
│                                TITLE                                      │
│ (  ) Draft        (  ) Final Environmental Statement                      │
│                                                                           │
│                                      Name, address, phone number of official │
│                                      who can be contacted for additional  │
│                                      information                          │
│                                                                           │
│ Responsible Office:                                                       │
│                                                                           │
│ 1. Name of Action: (  ) Administrative        (  ) Legislative           │
│                                                                           │
│ 2. Description of the Action: Brief summary of the proposed action and its purpose, highlighting important │
│    points. Areas affected - specific identity of states (and counties) particularly affected - maps. Other │
│    Federal actions in the area which are discussed in this EIS.           │
│ 3. Summary of Impacts                                                     │
│                                                                           │
│    a. Environmental Impacts: Brief summary of the impacts expected from the proposed action. │
│                                                                           │
│    b. Adverse Environmental Effects: Summary of the adverse environmental effects that will result if the │
│       proposed action is implemented.                                     │
│                                                                           │
│ 4. Alternatives: Brief discussion of the alternatives considered.        │
│                                                                           │
│ 5. List all Federal, state, and local agencies from which                │
│                                                                           │
│    a. Comments requested: for draft statements                           │
│                                                                           │
│    b. Comments received: for final statements, otherwise blank.          │
│                                                                           │
│ 6. Draft Statement to CEQ: Date the draft statement was forwarded to CEQ by ODUSA. │
│                                                                           │
│ 7. Final Statement to CEQ: Date the final statement was forwarded to CEQ by ODUSA. │
└─────────────────────────────────────────────────────────────────────────┘
```

Fig. 3.3 Summary sheet (prescribed by CEQ).

action with the necessary information to obtain an overview of the action and the circumstances surrounding it. Specifically, the project description should cover the following areas:

a. *Purpose of the action.* Describe the purpose of the action, and clearly state its goals and objectives. Explain what the action will accomplish.

b. *Description of the action.* Describe the proposal by name, and summarize the activities that will ensue if it is adopted. Some indication of the magnitude of the proposed action should be given, e.g., area extent, number of personnel involved, equipment, manpower, and material requirements.

c. *Environmental setting.* Describe the environment of the area affected as it exists prior to the proposed action. Both biophysical and socioeconomic aspects of the environmental setting should be included, e.g., location, mission, historical data, climate, topography, and population data. Unusual or important elements or features of the existing situation should be pointed out. For example, the existence of landmarks, rare timber stands, or unique community social characteristics should be identified.

1. PROJECT DESCRIPTION
   a. Purpose of action
   b. Description of action
      (1) Name
      (2) Summary of activities
   c. Environmental setting
      (1) Environment prior to proposed action
      (2) Other related Federal activities
2. LAND-USE RELATIONSHIPS
   a. Conformity or conflict with other land-use plans, policies and controls
      (1) Federal, state, and local
      (2) Clean Air Act and Federal Water Pollution Control Act Amendments of 1972
   b. Conflicts and/or inconsistent land-use plans
      (1) Extent of reconciliation
      (2) Reasons for proceeding with action
3. PROBABLE IMPACT OF THE PROPOSED ACTION ON THE ENVIRONMENT
   a. Positive and negative effects
      (1) National and international environment
      (2) Environmental factors
      (3) Impact of proposed action
   b. Direct and indirect consequences
      (1) Primary effects
      (2) Secondary effects
4. ALTERNATIVES TO THE PROPOSED ACTION
   a. Reasonable alternative actions
      (1) Those that might enhance environmental quality
      (2) Those that might avoid some or all adverse effects
   b. Analysis of alternatives
      (1) Benefits
      (2) Costs
      (3) Risks
5. PROBABLE ADVERSE ENVIRONMENTAL EFFECTS WHICH CANNOT BE AVOIDED
   a. Adverse and unavoidable impacts
   b. How avoidable adverse impacts will be mitigated
6. RELATIONSHIP BETWEEN LOCAL SHORT-TERM USES OF MAN'S ENVIRONMENT
   AND THE MAINTENANCE AND ENHANCEMENT OF LONG-TERM PRODUCTIVITY
   a. Trade-off between short-term environmental gains at expense of long-term losses
   b. Trade-off between long-term environmental gains at expense of short-term losses
   c. Extent to which proposed action forecloses future options
7. IRREVERSIBLE AND IRRETRIEVABLE COMMITMENTS OF RESOURCES
   a. Unavoidable impacts irreversibly curtailing the range of potential uses of the environment
      (1) Labor
      (2) Materials
      (3) Natural
      (4) Cultural
8. OTHER INTERESTS AND CONSIDERATIONS OF FEDERAL POLICY THAT OFFSET
   THE ADVERSE ENVIRONMENTAL EFFECTS OF THE PROPOSED ACTION
   a. Countervailing benefits of proposed action
   b. Countervailing benefits of alternatives

Fig. 3.4   Outline for CEQ-prescribed EIS content.

In developing the activity-description information, summary technical data, maps, and diagrams should be provided when relevant. Highly technical data may accompany the description as an attachment, or be footnoted and referenced, but should not appear in the description itself.

Information necessary to describe the environmental setting may be taken from either or both of the following sources:

- Existing data sources, e.g., soil, climatological, and hydrological information from master plans, USGS, SCS, EPA, state agencies, and historical societies;
- Acquired data sources, using available personnel capabilities or other means.

To ensure accurate descriptions and environmental assessments, site visits should be made when feasible. A first-hand accounting undoubtedly simplifies matters for both the preparer and the reviewer.

When population and growth characteristics are a factor, consideration should be given to using the rates of growth in the project region. These rates are contained in a projection compiled for the Water Resources Council by the Bureau of Economic Analysis of the Department of Commerce and the Economic Research Service of the Department of Agriculture (the "OBERS" projection).

All sources of data used to identify, quantify, or evaluate any and all environmental consequences must be expressly noted and referenced.

2. *Land Use Relationships.* Here, the description relates the proposed action to land use plans, policies, and controls for the affected area. Specifically, the discussion of land use relationships should cover the following areas:

a. *Conformity or conflict with other land use plans, policies, and controls.* Describe how the proposed action may conform or conflict with the objectives and specific terms of plans, policies, and controls imposed by either of the following categories:

(1) *Federal, state, and local.* Consider approved or proposed federal, state, and local land use plans, policies, and controls. Compare the land use aspects of the proposed action, and discuss possible compatibilities or conflicts. For example, discuss siting of an extremely noisy activity adjacent to a residential area, or the outleasing of land for purposes inconsistent with state wildlife management policies.

(2) *Clean Air Act or Federal Water Pollution Control Act Amendments of 1972.* This requires a discussion of policies and controls affecting land use plans developed in response to the Clean Air Act and the Federal Water Pollution Control Act of 1972. The extent of conformity or conflict with these land use plans must be considered. For example, the EPA has issued regulations requiring states to approve, in advance, the siting and con-

struction of both new "polluting facilities" and such "complex facilities" as shopping centers, amusement parks, and highways, which could cause a violation of air quality standards by attracting concentrations of vehicles.

b. *Conflicts and/or inconsistent land use plans.* When a conflict or inconsistency exists, the statement should describe the following:

    (1) *Extent of reconciliation.* Describe the extent to which the proponent has reconciled its proposed action with the plan, policy, or control, as described in part a above. For example, such a description may include statements indicating restriction of construction activities to normal daylight working hours to minimize noise impact, or restriction of agricultural activities on outleased land to protect wild life during the nesting season.

    (2) *Reasons for proceeding with action.* Notwithstanding the absence of full reconciliation, the proponent must explain the reasons why it has decided to proceed with the action.

3. *Probable Impact of the Proposed Action on the Environment.* Environmental consequences of the proposed action are summarized in this section. The assessment of the proposed action must include the following:

a. *Positive and negative effects.* Describe both beneficial and detrimental aspects of the environmental changes caused by the proposed action. Include commentary concerning the impacts on man's health, welfare, and surroundings. Considerations to be included are the following:

    (1) *National and international environment.* Consider the environmental consequences of actions, not only as they affect U.S. property and citizens, but as they affect areas in or under the jurisdiction of a nation other than the United States.

    (2) *Environmental factors.* Among factors to consider should be the potential effect of the action on such aspects of the environment as: air, water, land, ecology, sound, and socioeconomics.

    Details about these categories may be found in Appendix B.

    (3) *Impact of proposed action.* The attention given to different environmental factors will vary according to the *nature, scale,* and *location* of proposed actions. Primary attention should be given to discussing those factors most evidently impacted by the proposed action.

    (4) *Direct and indirect consequences.* Identify both direct and indirect effects of the proposed action, activity or project.

        • *Primary effects.* Include direct impacts on man's health and welfare and on other forms of life and related ecosystems. Examples of direct effects might include noise from aircraft operations or benefits from installation of wet scrubbers to meet air quality standards.

- *Secondary effects.* Include secondary and indirect environmental impacts, particularly on population concentration and growth. Many federal actions attract people to previously unpopulated areas, and indirectly cause pollution, congestion, and land development that probably would not have existed otherwise. Conversely, other actions may result in displacing population. As an example, noise from aircraft operations may affect future land use patterns in the area, and the results of air pollution abatement operations may produce a secondary water pollution problem from the scrubber waste.

4. *Alternatives to the Proposed Action.* The National Environmental Policy Act specifically requires the proponent to "study, develop, and describe appropriate alternatives to recommended courses of action in any proposal which involves unresolved conflicts concerning alternative uses of available resources." Point 4 responds to this requirement.

Alternatives to the proposed action, including (when relevant) those not within the existing authority of the responsive agency, should be discussed. The purpose of environmental "pros and cons" for alternatives may be examined early in the planning process in order not to prematurely foreclose options which might enhance environmental quality or have reduced detrimental effects.

When a preferred course of action is assessed, and when it is determined that there is no significant environmental impact, there is normally no need to assess alternative courses of action. On the other hand, when there are several courses of action which are equally acceptable, it is appropriate to attempt to identify the alternative which will have the least environmental impact.

a. *Reasonable alternative actions.* In addressing this point, it is necessary to provide a rigorous exploration and objective evaluation of the environmental impacts of all *reasonable* alternative actions, particularly:

(1) Those which might enhance environmental quality, and

(2) Those that might avoid some or all adverse environmental effects.

b. *Analysis of alternatives.* When evaluating alternatives, the following consequences should be considered and analyzed:

*Refer to Secondary Impacts section in Chapter 4.

(1) *Benefits.* Discuss benefits which may be derived from an alternative (specifically, single out those specified in part a above).

(2) *Costs.* Give an accounting of the environmental costs involved in each alternative, so that tradeoffs between environmental protection and enhancement and associated costs may be evaluated.

(3) *Risks.* Include specific details regarding potential adverse effects of each alternative, so that these consequences may be weighed against the project's costs and benefits. When discussing alternatives to the proposed action, the following examples of alternative possibilities should be considered:

- No action. Treat the alternative of taking no action with the same analysis as outlined above (specifically required).
- Rescheduling action. Consider the effects of delaying the action (e.g., postponing temporary habitat disruptive activities until after the nesting season, or delaying clearing operations until after the wet season).
- Plan modification. Examine alternative means of accomplishing the mission or action which would provide similar benefits with different environmental impacts (e.g., simulated versus live firing or testing, or temporary versus permanent haul roads).
- Different design and/or site location. Consider alternatives related to different designs or details which would present different environmental impacts (e.g., cooling ponds versus cooling tower for a power plant, or alternatives that would significantly conserve energy).
- Compensatory alternatives. Examine alternatives that may compensate for resource or other losses. For example, the acquisition of land, waters, and interests therein to compensate for the loss of fish and wildlife habitat caused by the proposed project.

5. *Probable Adverse Environmental Effects Which Cannot be Avoided.* This section should be a brief summary of the effects discussed under Point 3 that are adverse and unavoidable under the proposed action. In response to this point, the following specific items are required:

a. *Adverse and unavoidable impacts.* Summarize those probable adverse effects which cannot be avoided if the proposal were to be implemented. In addition to an evaluation of damage to the natural environment, this would include a summary evaluating the extent to which human health or safety, aesthetically or culturally valuable surroundings, standards of living, and other aspects of life, would be sacrificed or endangered. Such impacts may include the following:

- Water or air pollution resulting from the action;
- Destruction of historical or archaeological sites;
- Disruption of wildlife habitats;
- Increase in urban congestion;
- Threats to health;
- Undesirable land use patterns;
- Other consequences adverse to the environmental goals of the National Environmental Policy Act.

  b. *Avoiding Adverse Impacts.* For purposes of contrast, include a clear statement of how adverse effects previously discussed will be mitigated. For example, the following mitigation statements might be applicable:

  - Wildlife and vegetation will be protected by limiting vehicular traffic related to construction activities to those areas which will be disrupted by permanent construction.
  - Advance coordination with local public school officials will allow for sufficient planning to accommodate anticipated increased enrollment pressures on local public schools.
  - Land areas will be sodded after completion of site clearing, to prevent unnecessary erosion.
  - Haul roads will be treated with palliatives to minimize dust production.

  From the above examples, it is clear that if the mitigation of impacts section is to have any meaning, the mitigation procedures indicated in the assessment or statement must be carried out during implementation of the proposed action. Thus, the inclusion of specific mitigation procedures must be coordinated closely with the proponent, and must become a part of the overall design or incorporated into the plan for the proposed action.

6. *Relationships Between Local Short-Term Uses of Man's Environment and the Maintenance and Enhancement of Long-Term Productivity.* This section should contain a brief discussion of the extent to which the proposed action involves short-term versus long-term environmental gains and losses. In this context, short-term and long-term do not refer to any fixed time periods, but should be viewed in terms of the environmentally significant consequences of the proposed action. Thus, "short-term" may range from a very short time during which an action takes place, to the expected *life* of a project. Cumulative effects should be discussed in this section. Some of the impacts discussed in Point 3 may be repeated (however, they should be placed in long-term versus short-term format).

a. *Trade-off between short-term environmental gains at expense of long-term losses.* Discuss the relationship between short-term, beneficial environmental effects, and loss or decrease in long-term productivity or use of resources such as land. Incorporate into the discussion any cumulative effects caused by continuing or repeated activities. For example, application of herbicides or pesticides may remove undesirable species, but long lasting or cumulative effects may permanently damage other vegetative growth or result in disruption of ecological balance.

b. *Trade-off between long-term* environmental gains at expense of *short-term losses.* Discuss positive aspects of the proposed action on a long-term basis and compare these with detrimental short-term effects. For example, construction of a sewage treatment plant may result in activities which create noise, dust, or erosion, but long-term aspects of the project include enhanced water quality in the receiving stream.

c. *Extent to which proposed action forecloses future options.* Assess the cumulative and long-term impacts of the proposed action with the view that each generation is a trustee of the environment for succeeding generations. Consider such losses as restrictions on visitations of historical or archaeological sites, destruction of natural vistas, or increased danger to threatened species.

7. *Irreversible and Irretrievable Commitments of Resources.* This section identifies irrevocable effects on resources that would be involved in the proposed action if it were implemented. This is accomplished by reviewing the discussion of Point 5, and by specifically identifying *unavoidable impacts that irreversibly curtail the range of potential uses of the environment.*

When discussing resources committed to loss or destruction by the action, the term "resources" should include the following meanings:

a. *Labor.* Discuss the labor requirements for the proposed action and the degree to which the expenditure of this labor force would detract from other areas of productivity.

b. *Materials.* Discuss the use of materials in short supply (e.g., fuels, wood and scarce metals), but do not include materials which are plentiful or have competitive alternatives (e.g., aggregate, fill material).

c. *Natural.* Discuss the irrevocable use of natural resources resulting in effects such as ecosystem imbalance, destruction of wildlife habitats, or loss of natural land use patterns. Specifically include consumption of natural energy resources in short supply, such as oil or natural gas.

   d.   *Cultural.* Discuss destruction of human interest sites, archaeological sites, scenic views, or open space where such space is limited. Reiterate lasting social or economic effects that the proposed action might have on the surrounding community.

8. *Other Interests and Considerations of Federal Policy That Offset the Adverse Environmental Effects of the Proposed Action.* This section is designed specifically to call attention to positive aspects of the proposed action that might offset the negative effects revealed in the discussion of Points 3 and 5. The benefits of the proposed action should be addressed in the following manner:

   a.   *Countervailing benefits of the proposed action.* Discuss beneficial aspects of the action, in order that positive effects may be weighed against negative environmental effects.

   b.   *Countervailing benefits of alternatives.* Discuss the extent to which the benefits stated in part (a) could be realized by following reasonable alternatives (identified in Point 4) that would avoid some or all of the adverse environmental effects. If a cost-benefit analysis has been prepared for the proposed action, it should be attached. However, the extent to which environmental costs have or have not been reflected in the analysis should be clearly defined.

3.6 PROCESSING OF EIA/EIS

After the environmental impact assessment or statement has been prepared, processing of the document depends upon whether the document is (1) an assessment; (2) a negative declaration statement; (3) a draft statement; or (4) a final statement.

Assessment. If an environmental assessment is made, and it is determined that no statement is required, the documentation, processing, and other follow-up procedures may vary both between and within agencies. In many cases, the proponent of the action is required only to document the assessment and retain a copy in the project files. Other cases may require that copies be forwarded to offices within the agency, as specified in guidance issued by the proponent agency.

Negative Declaration Statements. If an agency determined that an environmental impact statement is not necessary, but that a negative declaration statement will be required, the agency is required to prepare a publicly available record, briefly setting forth the agency's decision and the reasons for the negative determination. Agencies are to maintain, periodically revise, and make available for public inspection on request, lists of such negative determinations. In addition, the lists are to be forwarded to CEQ, which will periodically publish such lists in the *Federal Register*.

Draft Statements. If it is determined that an environmental impact statement is required, the next step is the preparation and processing of a *draft* environmental impact statement (see Fig. 3.5). The document normally is prepared according to guidelines issued by the proponent agency; after undergoing internal review (which again may vary between or within agencies), copies are sent to CEQ, and the draft statement undergoes a review process outlined by CEQ.[4] Specifically, the review process includes the following:

1. *Federal agency review.* The proponent agency should consult with and obtain specific comments on the draft statement from other federal and federal-state agencies with jurisdiction by law or special expertise in specific areas of environmental impact. Specific federal and federal-state agencies and their corresponding areas of expertise are listed in Appendices II and III of the CEQ Guidelines.[4] These comments are to be pursued *in*

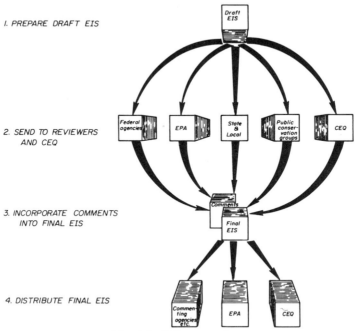

Fig. 3.5   EIS processing.

addition to other specific statutory obligations requiring counsel or coordination with other federal or state agencies (e.g., resulting from the Fish and Wildlife Coordination Act, National Historic Preservation Act of 1966, etc.).

2. *EPA review.* If the proposed project is related to provisions of the authority of the Administrator of the Environmental Protection Agency (i.e., air quality, water quality, pesticide regulation, noise abatement and control, solid waste disposal, and environmental radiation criteria and standards), the proponent agency must submit the environmental impact statement to the Administrator for review and written comment. EPA, in turn, determines whether (1) the statement is adequate; (2) the statement is inadequate; or (3) the project is environmentally unsatisfactory. EPA then publishes its determination, and notifies CEQ if the statement is found to be inadequate or the project is found to be environmentally unsatisfactory. Normally, an attempt is made first to obtain a revision of the statement or project, if either one does not meet the necessary standards.

3. *State and local review.* State and local agencies which can assist in preparation and review of environmental statements must be included in draft statement processing. A system of state and area clearinghouses of the Office of Management and Budget (OMB) provides a means for obtaining this state and local review. Appendix IV of the CEQ Guidelines contains a list of instructions for carrying out the established procedures.[4]

4. *Public review.* Public participation in the review process is also encouraged, although the specific procedures for obtaining public review is the responsibility of the agencies. Usually, this is accomplished by publishing newspaper notices regarding the availability of the draft statements; by holding public hearings; and by maintaining lists of interested conservation groups and individuals and providing them with project information and copies of the draft statement.

The actual time allotted for the review process is dependent upon the magnitude and complexity of the statement and the extent of citizen interest in the proposed action. A minimum of 45 days is required for the review period, and reviewer requests for extensions of time to complete the review process may be granted.

Final Statements. After reviewing the comments from the review process, the agency should then make every effort to address substantive comments, issues, and views brought to the agency's attention, and incorporate the responses into the final statement. After agency procedures have been satisfied, copies of the final environmental statement, together with the substance of all comments received from the review process, are forwarded to CEQ. In addition, copies of final statements, with attached review comments, are sent to all reviewers who provided comments, and to the EPA for assistance in carrying out its responsibilities under the Clean Air Act.

Time Requirements. CEQ has requested that "to the maximum extent practicable," there should be 90 days between issuance of a draft statement and any administrative action on the project.[4] Furthermore, there should be 30 days between issuance of the final statement and the beginning of administrative action. This is to ensure that sufficient time is provided for the careful review of environmental considerations. When the final statement is filed *within* 90 days after the draft statement has been circulated for comment, furnished to CEQ, and made public, the minimum 30- and 90-day periods may overlap.[4]

REFERENCES

1. *Environmental Quality*, Third Annual Report of the Council on Environmental Quality, U.S. Government Printing Office (USGPO), August 1972.
2. *Environmental Quality*, Fourth Annual Report of the Council on Environmental Quality, USGPO, September 1973.
3. National Environmental Policy Act of 1969 (PL 91-190; 83 Stat 852).
4. "Preparation of Environmental Impact Statements: Guidelines," *Federal Register* **38,** *No. 147*, Part II: 20550–20562 (August 1, 1973).

4

Elements of Environmental Impact Analysis

As was indicated in previous chapters, environmental impact analysis encompasses varied disciplines, and consequently, requires the expertise of personnel knowledgeable in the various technical areas. When assessing the environmental impact of a given project, four major elements are involved:

1. Determining agency activities associated with implementing the action or the project;
2. Determining environmental attributes (elements) representing a categorization of the environment such that changes in the attributes reflect impacts;
3. Determining environmental impact; and
4. Reporting findings.

4.1 AGENCY ACTIVITIES

A comprehensive list of activities associated with implementing the project or action throughout its life cycle should be developed. Necessary levels of detail would depend upon the size and type of project. As an illustration, an example of detailed activities for construction are included in the matrix in Appendix C.

4.2 ENVIRONMENTAL ATTRIBUTES

Consisting of both natural and man-made factors, the environment is difficult to characterize because of its many attributes (elements) and the complex interrelationships among these attributes. Changes in the attributes of the environment and their interrelationships are defined as impacts.

An EIA or EIS is prepared to characterize the environment and potential changes to be brought about by a specific activity. Such a document is advantageous in that it presents an organized and complete information base for achieving the benefits intended by NEPA. In order for this objective to become fulfilled, it is necessary that a complete description and, hence, *understanding* of the environment to be affected is first achieved. Various impact assessment methodologies have been developed (see Chapter 5), and virtually all of them employ a categorization of environmental characteristics in some form or other. This approach is recommended in order that aspects of the environment are not overlooked during the analysis phase.

Definition: Variables that represent characteristics of the environment are defined as attributes, and changes in environmental attributes provide indicators of changes in the environment.   All lists of environmental attributes are a shorthand method for focusing on important characteristics of the environment. Due to the complex nature of the environment, it should be recognized that any such listing is limited and, consequently, may not capture some of the real world impacts. The more complete the listing is, the more likely it will reflect all important effects on the environment, but this may be expensive and cumbersome to apply.

Figure 4.1 presents a general listing of 49 attributes in seven categories encompassing the biophysical and socioeconomic environment. While it is felt that this list of attributes represents a reasonable, concise breakdown of environmental parameters, depending upon the type of action to be assessed, this list may need to be modified or supplemented. Appendix B relates details of the specific attributes, and the following sections provide a general discussion of the seven categories.

Air

When assessing the primary resources that are needed to sustain life, one must consider air as being one of the most, if not *the* most, critical resource.

Air
1. Diffusion factor
2. Particulates
3. Sulfur oxides
4. Hydrocarbons
5. Nitrogen oxide
6. Carbon monoxide
7. Photochemical oxidants
8. Hazardous toxicants
9. Odor

Water
10. Aquifer safe yield
11. Flow variations
12. Oil
13. Radioactivity
14. Suspended solids
15. Thermal pollution
16. Acid and alkali
17. Biochemical oxygen demand (BOD)
18. Dissolved oxygen (DO)
19. Dissolved solids
20. Nutrients
21. Toxic compounds
22. Aquatic life
23. Fecal coliform

Land
24. Soil stability
25. Natural hazard
26. Land use patterns

Ecology
27. Large animals (wild and domestic)
28. Predatory birds
29. Small game
30. Fish, shell fish, and water fowl
31. Field crops
32. Threatened species
33. Natural land vegetation
34. Aquatic plants

Sound
35. Physiological effects
36. Psychological effects
37. Communication effects
38. Performance effects
39. Social behavior effects

Human Aspects
40. Life styles
41. Psychological needs
42. Physiological systems
43. Community needs

Economics
44. Regional economic stability
45. Public sector review
46. Per capita consumption

Resources
47. Fuel resources
48. Nonfuel resources
49. Aesthetics

Fig. 4.1   Environmental attributes.

What makes air quality particularly vulnerable is that air, unlike water or other wastes, cannot be reprocessed practically at some central location and subsequently distributed for use. If the air becomes poisonous, the only alternative to sustain life is for each individual to wear some sort of life support system. This is clearly unworkable and economically infeasible. When emissions and unfavorable climatic conditions interact to create undesirable air quality, the atmospheric environment may begin to exert adverse effects on man and his surroundings. Air may be replenished through photosynthetic processes and cleansed through precipitation, but these natural processes are limited in their effectiveness. Hence, great care must be exercised when assessing and maintaining the quality of air resources. It therefore seems self evident that the protection of our air quality is a vital consideration when assessing the environmental impact of man's diversified activities.

To better understand why our air quality has deteriorated and will probably continue to deteriorate, even if the most advanced technology developed to date is applied, one must recognize the factors responsible for air pollution problems. Air quality is intimately connected to population growth, expansion of industry and technology, and urbanization. In particular, the energy use associated with these activities is increasing at a rate that may double energy consumption in the next 25 years. Since energy use and air pollution are very strongly correlated, it seems imperative that we, as a society, examine each of our everyday activities in light of its potential impact on the environment. In effect, we must examine our life style, both at a professional and a personal level, to assure that the precious resource, clean air, is preserved.

The Clean Air Act of 1970 was established "to protect and enhance the quality of the nation's air resources so as to promote public health and welfare and the productive capacity of its population." In 1971, the EPA set forth national primary and secondary ambient air quality standards under Section 109 of the Clean Air Act. The ambient standards are shown in Table 4.1. The primary standards define levels of air quality necessary to protect public health, while secondary standards define levels necessary to protect the public welfare from any known or anticipated adverse effects of a pollutant.

In addition, the EPA established emission standards for new stationary sources. Unless state or local standards are more stringent, these regulations apply to any new (effective 17 August 1971) building, structure, facility, or installation which emits or may emit any air pollutant. The standards, summarized in Table 4.2, are also applicable to modifications in existing plants.

For new and existing federal facilities and buildings not covered by the EPA Regulation discussed previously, the particulate emissions are limited to those shown in Fig. 4.2.

The pollutants identified in the EPA ambient standards each have different effects on the health and welfare of man, as summarized in Table 4.3.

Transportation and fuel combustion in stationary sources contribute highly to air pollution in the United States (see Table 4.4). Other activities involved with industrial operations and solid waste disposal also add a substantial amount of pollutants.

To assess the impact of various activities on the air quality, the major elements of the air pollution problem may be examined. These are (1) the presence of a source; (2) a means of transporting the pollutant to a receptor; and (3) the receptor. If any of these elements are removed, the problem ceases to exist. When examining sources, two types of classifications may be used: particulates, and gases and vapors. Under the particulate category, one finds smoke, dust, and fumes, as well as liquid mists. To further identify the impact of these particulates, it may be necessary to further subdivide into chemical and biological classifications. Likewise, for gases and vapors

TABLE 4.1   Ambient Air Quality Standards
(EPA Regulation—40 CFR 50; 36 FR 22384, 25 November
1971)

Pollutant	Ambient Standards
Carbon monoxide (CO)	Primary and secondary standards are 10 milligrams per cubic meter (9 ppm) as a maximum eight-hour concentration not to be exceeded more than once a year, and 40 milligrams per cubic meter (35 ppm) as a maximum one-hour concentration not to be exceeded more than once a year.
Photochemical oxidants	Primary and secondary standards are 160 micrograms per cubic meter (.08 ppm) as a maximum one-hour concentration not to be exceeded more than once a year.
Hydrocarbons (HC)	Primary and secondary standards are 160 micrograms per cubic meter (.24 ppm) as a maximum three-hour concentration (6 to 9 a.m.) not to be exceeded more than once a year.
Nitrogen oxides (NOx)	Primary and secondary standards are 100 micrograms per cubic meter (.05 ppm) on an annual arithmetic mean.
Sulfur oxides (SOx)	Primary standard is 80 micrograms per cubic meter (.03 ppm) on an annual arithmetic mean, and 365 micrograms per cubic meter (.14 ppm) as a maximum 24-hour concentration not to be exceeded more than once a year. The secondary standard is 60 micrograms per cubic meter (.02 ppm) on an annual arithmetic mean, 260 micrograms per cubic meter (.1 ppm) maximum 24-hour concentration not to be exceeded more than once a year, and 1,300 micrograms per cubic meter (.5 ppm) as a maximum three-hour concentration not to be exceeded more than once a year.
Particulate matter	Primary standard is 75 micrograms per cubic meter on an annual geometric mean, and 260 micrograms per cubic meter as a maximum 24-hour concentration not to be exceeded more than once a year. The secondary standard is 60 micrograms per cubic meter on an annual geometric mean, and 150 micrograms per cubic meter as a maximum 24-hour concentration not to be exceeded more than once a year.

TABLE 4.2   EPA Emmission Standards for New Stationary Sources (40 CFR 60; 36 FR 24876, 23 December 1971; Effective 17 August 1971)

Source	Emission Standards		
	Particulates	Sulfur Dioxide	Nitrogen Oxides
Fossil fuel fired steam generator (250 million BTU/hr input size and up)	No discharge shall be: (a) In excess of 0.10 lb per million BTU heat input (0.18 g per million cal), maximum 2-hour average. (b) Greater than 20 percent opacity, except that 40 percent opacity shall be permissible for not more than 2 minutes in any hour. (c) Where the presence of uncombined water is the only reason for failure to meet the requirements of paragraph (b) of this section such failure shall not be a violation of this section.	No discharge shall be in excess of: (a) 0.80 lb per million BTU heat input (1.4 g per million cal), maximum 2-hour average, when liquid fossil fuel is burned. (b) 1.2 lbs per million BTU heat input (2.2 g per million cal), maximum 2-hour average, when solid fossil fuel is burned. (c) Where different fossil fuels are burned simultaneously in any combination, the applicable standard shall be determined by proration.	No discharge shall be in excess of: (a) 0.20 lb per million BTU heat input (0.36 g per million cal), maximum 2-hour average, expressed as NO_2, when gaseous fossil fuel is burned. (b) 0.30 lb per million BTU heat input (0.54 g per million cal), maximum 2-hour average expressed as NO_2, when liquid fossil fuel is burned. (c) 0.700 lb per million BTU heat input (1.26 g per million cal). (d) When different fossil fuels are burned simultaneously in any combination the applicable standard shall be determined by proration.
Incinerator (50 ton/day charging rate and up)	No discharge shall be in excess of 0.08 gr/scf (0.18 g/NM^3) corrected to 12% CO_2, maximum 2-hour average.	N/A	N/A
Portland cement plants	No kiln discharge shall be: (a) In excess of 0.30 lb per ton of feed to the kiln (0.15 Kg per metric ton), maximum 2-hour average. (b) Greater than 10 percent opacity, except that where the	N/A	N/A

... is the only reason for failure to meet the requirements for this subparagraph, such failure shall not be a violation of this section.

No clinker cooler discharge shall be:
(1) In excess of 0.10 lb per ton of feed to the kiln (0.050 Kg per metric ton), maximum 2-hour average.
(2) 10 percent opacity or greater.
No discharge from any affected facility other than the kiln and clinker cooler shall be 10 percent opacity or greater.

Facility			
Nitric acid plants	N/A	N/A	No discharge shall be: (a) In excess of 3 lbs per ton of acid produced (1.5 kg per metric ton), maximum 2-hour average, expressed as NO_2. (b) 10 percent opacity or greater.
Sulfuric acid plants	N/A	No discharge shall be in excess of 4 lbs per ton of acid produced (2 kg per metric ton), maximum 2-hour average. No acid mist discharge shall be: (a) In excess of 0.15 lb per ton of acid produced (0.075 kg per metric ton), maximum 2-hour average, expressed as H_2SO_4. (b) Nitrogen oxides 10 percent or greater.	N/A

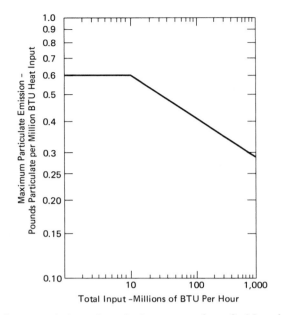

Fig. 4.2   Maximum emission of particulate matter from fuel-burning installations for both new and existing federal facilities and buildings (40 CFR 76; 36 FR 22417, 25 November 1971).   Emission standards for five specific types of new stationary sources are shown in Table 4.2.

one may consider sulphur oxides, nitrogen oxide, carbon monoxide, hydrocarbons, and hazardous toxicants.

Finally, the environmental influence factors affect the transport mechanism of the pollutant. The pollutant transport, or lack of it, is controlled by meteorological and topographical conditions. Clearly, less ground level pollutant concentration will occur on a flat, open plain under windy conditions than in a valley under calm conditions. These factors and situations are discussed for the Air attributes listed in Appendix B.

The receptors for air quality impact are not humans alone. Air pollution has been definitely identified as having deleterious effects on humans, animals, plant life, and materials. A drastic reduction in air quality is bound to severely affect the overall ecosystem behavior.

Water

Water of high quality is essential to human life, and water of acceptable quality is essential for agricultural, industrial, domestic, and commercial uses; in addition, most recreation is water-based; therefore, major activities

TABLE 4.3 Effects of Some Major Air Pollutants on Man

Pollutant	Effect
Carbon monoxide	Combines with hemoglobin in blood, displacing the vital oxygen that hemoglobin normally transports, thereby reducing the oxygen-carrying capacity of the circulatory system. Results in reduced reaction time and increased burden on pulmonary system in cardiac patients.
Photochemical oxidants	Reacts with nitrogen oxides and hydrocarbons in the presence of sunlight to form photochemical smog; causes eye, ear, and nose irritation and adversely affects plant life.
Hydrocarbons	Combine with oxygen and NOx to form photochemical oxidants.
Nitrogen oxides	Forms photochemical smog and is associated with a variety of respiratory diseases.
Sulfur oxides	Associated with respiratory diseases and can form compounds resulting in corrosion and plant damage.
Particulate matter	Injures surface within respiratory system, causes pulmonary disorders and eye irritation, and creates psychological stress. Results in economic loss from surface material damage.

TABLE 4.4 Estimated Emissions of Air Pollutants by Weight, Nationwide, 1971[8] (Millions of Tons per Year)

Source	CO	Particulates	SO_2	HC	NO_x
Transportation	77.5	1.0	1.0	14.7	11.2
Fuel combustion in stationary sources	1.0	6.5	26.3	.3	10.2
Industrial processes	11.4	13.6	5.1	5.6	.2
Solid waste disposal	3.8	.7	.1	1.0	.2
Miscellaneous	6.5	5.2	.1	5.0	.2
Total	**100.2**	**27.0**	**32.6**	**26.6**	**22.0**
Percent change 1970 to 1971	−.5	+5.9	−2.4	−2.6	0

The table does not include data on photochemical oxidants because they are secondary pollutants formed by the action of sunlight on nitrogen oxides and hydrocarbons and thus are not emitted from sources on the ground.

having potential effects on surface water are certain to be of appreciable concern to the consumers and taxpayers who support these activities. Additionally, developments of recent years suggest that Americans are far more concerned about water quality than in previous years.

Perhaps the political process provides the best barometer to measure the extent of public concern about water quality. The United States House of Representatives and Senate overwhelmingly enacted (over Presidential veto) the Federal Water Pollution Control Act Amendments of 1972. This legislation, in general, calls for uniform application of "secondary treatment" by 1 July 1977, at publicly owned treatment works, and "application of best practicable control technology currently available for other point discharges." By 1 July 1983, point discharges other than publicly owned treatment works "shall require application of the best available technology economically achievable for such category or class, which will result in reasonable further progress toward the national goal of eliminating the discharge of all pollutants."

Additionally, and significantly, the Act requires that agencies of the federal government comply with federal, state, interstate, and local requirements respecting control and abatement of pollution to the same extent that any person is subject to such requirements. Potential impacts on surface water quality and quantity are certain to be of concern in assessment of the effects of many federal programs. Almost any activity of man offers the potential for impact on surface water through generation of waterborne wastes, alteration of the quantity and/or quality of surface runoff, direct alteration of the water body, modification of the exchanges between surface and groundwaters through direct or indirect consumption of surface water, or other causes.

The hydrologic environment is composed of two interrelated phases: groundwater and surface water. Impacts initiated in one phase eventually affect the other. For example, a groundwater system may charge one surface water system and later be recharged by another surface water system. The complete assessment of an impact dictates consideration of both groundwater and surface water. Thus, pollution at one point in the system can be passed throughout, and consideration of only one phase does not characterize the entire problem.

Due to the close interrelationship between surface and groundwaters, most environmental attributes inevitably interface. Hence, aside from those aspects dealing specifically with surface or subterrain features, the attributes may be considered as applicable to both.

Many attributes of the aquatic environment could be categorized as being physical, chemical, or biological in nature. *Physical attributes* of surface water could be categorized as relating to either the physical nature of the water body or to the physical properties of the water contained therein. Examples of individual parameters in the former category would include the depth, velocity, and rate of discharge of a stream. Features of this type might be influenced by major activities, such as withdrawal of water, dredging, and clearing of vegetation. The other category of physical characteristics—those related to the water itself—includes water characteristics such as color, turbidity, temperature, and floating solids. Many types of activities could influence the physical properties of water. A few examples are clearing of land and construction of hardstands, roads and rooftops (which might accelerate erosion, flooding, and sedimentation), discharge of scale-laden boiler waters, and discharge of cooling waters. Some other quality aspects which could be included in this category are dissolved gases and tastes and odors, which are actually manifestations of chemical properties of water. This serves to illustrate the occasional difficulty in categorizing attributes of the water environment.

Chemical attributes could be categorized conveniently as organic or inorganic chemicals. Some inorganic chemicals (like cadmium, lead, and mercury) may have grave consequences to human health; some (notably phosphorus and dissolved oxygen) have severe effects on the water environment, while others (such as calcium, manganese, and chlorides) relate mainly to man's economic and aesthetic value of water in his commercial, industrial, and domestic uses. Normal personal use of water by man increases the concentration of many inorganic chemicals in water. Additionally, almost any type of industrial activity and land drainage is a source. Because of the hundreds of thousands of organic (carbon-based) chemicals produced naturally and by man, most of the attributes contained in the organic chemical category are "lumped-parameters." Examples include BOD (biochemical oxygen demand), oil, and toxic compounds. Some organic compounds are natural constituents of surface drainage and human and animal wastes, while others are unique to industrial activities and industrial products.

Biological attributes of the water environment conveniently could be categorized as pathogenic agents or aquatic life. Pathogenic (disease-causing) agents include certain virus, bacteria, protozoa, and other organisms, and they originate almost exclusively from human wastes. Aquatic life refers to the microorganisms and microscopic plants and animals, including fish, which inhabit water bodies. They are affected directly or indirectly by almost any natural or man-made change in a water body.

It is difficult to conceive of an alteration of surface water quantity or quality which is not accompanied by secondary effects. The physical, biological, and chemical factors influencing water quality are so interrelated that a change in any water quality parameter triggers other changes in a complex network of interrelated variables. Thus, while individual water quality and quantity parameters may seem far more amenable to quantitative expression than parameters describing the quality of other sectors of the environment, the total effect of a particular impact on surface water may be as intangible as those on any sector of the total environment, because of the complex secondary, tertiary, and higher order effects.

Land

Considering both the physical makeup and the uses to which it is put, land constitutes another important category of the environment. The soil that mantles the land surface is the sole means of support for virtually all terrestrial life. As this layer is depleted by improper use, so is the buffer between nourishment and starvation destroyed. However, the ability of soil to support life varies from place to place according to the nature of the local climate, the surface configuration of the land, the kind of bedrock, and even the type of vegetation cover. At the same time, the vulnerability of soil to destruction through mismanagement will vary as these factors change. Cultivated soils on slopes greater than 6 percent, or those that developed on limestone, are prone to erosion; soils in arid climates are sensitive to degradation by excessive salt. On the other hand, those in the tropics may quickly lose their plant nutrients by exposure to the abundant rainfall of those areas.

Soil serves well as an example of an interface between the three great systems that comprise the earth sciences: the lithosphere, the atmosphere, and the hydrosphere. The biosphere also operates in this interface, but it is usually considered to comprise the life sciences. For purposes of this discussion, the lithosphere consists of the various characteristics of landforms (slope, elevation, etc.), landform constituent materials (substratum), and the weathered layer or soil regolith. In the case of the atmosphere, the main elements are those that describe its state of temperature, moisture and motion—or, in a word, climate. With regard to the hydrosphere, the principal concern will be with water flowing over the land surface in the form of sheet wash and streams.

Climate profoundly influences the nature and site characteristics, such as soils and vegetation. Soil trafficability, to a substantial degree, is the result of the interaction of rainfall and temperature with the local rock types. The rate of soil erosion, other things being equal, will depend upon the amount and intensity of rainfall. The details of the site climate must be known before an adequate environmental impact assessment can be made.

Climates are commonly identified and described by the total annual amount of precipitation and its seasonal distribution, and by temperature and its seasonal distribution. Climatic types may be described as warm-humid, cool-humid, cool with summer droughts, arid, semi-arid, and so on. There are additional descriptive elements of climate that are important in causing substantial differences within any one climatic type. Some of these include probability of maximum rainfall intensity, probability of drought, length of growing season, wind intensity, and the kind and frequency of storms.

The preparer of an environmental impact assessment or statement should be aware of the local landform type and its constituent materials. This information will enable him to more quickly evaluate the potential hazards of his activity upon the local physical environment. For example, slope erosion problems should be slight in plains areas with low relief. Areas underlain by limestone must always be treated cautiously, with respect to groundwater pollution.

For purposes of description, the schematic graph in Fig. 4.3 will permit a quick identification of basic terrain types from topographic maps. The

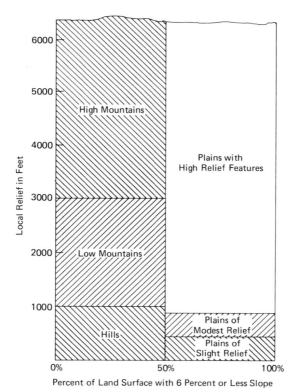

Fig. 4.3   Terrain types of the world.[2]

more common types of landform-constituent materials can usually be identified fairly easily in the field.

This initial breakdown of landform types is based upon only two descriptive characteristics of topography: local relief and slope. Other important properties are: pattern, texture, constituent material, and elevation. These, along with local relief and slope, can be used to identify landforms with a considerable degree of precision.

However, the above define the landform system only at a given moment of time. Landforms are not static, but are continually changing; i.e., the landform system is dynamic, since landforming processes are continuously at work, although the rate at which they operate varies from place to place. The factors that influence process rate include some of the attributes of landforms, as well as the attributes of climate and biota. Fig. 4.4 shows one way of illustrating this complex system.

It is evident, from the above relationships, that landform evolution can also be considered important. Among the more important processes are weathering (disintegration and decay of rock), stream and wind erosion (removal of weathered debris by those forces), mass wasting (direct removal of weathered material by gravity), deposition (the cessation of movement of the entrained rock debris), and soil formation (those processes of weathering that give soils their distinctive regional characteristics). It is also evi-

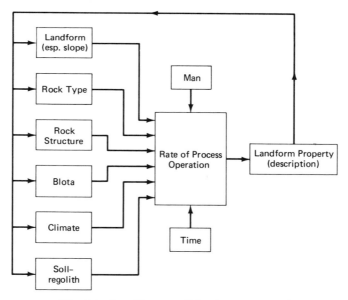

Fig. 4.4   Landform modification processes.[2]

dent that man is an important factor in changing the rate of process operation. He does this by modifying the land surface—by changing the vegetation or destroying it; by plowing or otherwise disturbing the soil-regolith; by paving, construction, or otherwise sealing the surface; by changing the chemical or physical equilibrium in the soil-regolith, and so forth. These and other actions man takes reduce the natural resistances of the physical environmental system and permit the physical processes to operate at accelerated rates—not only with respect to one attribute, but to many.

As a typical historical example,[2] one might consider the Piedmont region of the southeastern United States, where the interaction of soils, slope, climate, and the substitution of clear field cultivation of cotton and tobacco led to widespread destruction of the physical setting. The bare, gentle-to-moderate slopes, combined with the clay-rich middle layers of the local soils, and the extremely heavy late summer and fall precipitation associated with hurricanes, created circumstances of exceedingly rapid soil erosion. The intense runoff quickly formed gullies in the surface layers and commenced to meander laterally, stripping off the soil. Once gullies were eroded through the mid layers, they deepened and lengthened rapidly, the water table was lowered, and the potential plant growth was thereby diminished. The process continued, with damage spreading to all parts of the system, eventually returning to man with a vengeance. The vegetation was impoverished, the wildlife destroyed, and the streams were polluted with excessive sediment. This process was advanced enough that land abandonment in the Piedmont began as long ago as the beginning of the nineteenth century.

All social and economic activities are located in time and space. The spatial or locational aspects of these activities involve land in some way. Thus, land is a resource; i.e., it is useful in the production of goods and services needed to satisfy human wants and desires.

Land may be used directly, as in agriculture or forestry, where production depends partly on the inherent capability of the soil, and where land serves to locate the activity in space. Or, land may mainly provide the locational base upon which all sorts of commercial structures, transportation, and communication facilities or residential housing are built, and on which all sorts of social and economic activities take place.

Man's activities mainly affect the availability or suitability of land for certain uses and thus, land patterns. The activities may have negative or or positive repercussions of varying magnitude on the local or regional economy, or on community social or cultural patterns or on the biophysical characteristics of the land itself, depending on the nature and extent of the activity. For instance, increases in personnel, due to a major federal action, may cause shortages of presently available rental housing, as well as cause rent increases. However, increased housing demand may stimulate residential and related construction requiring more land, thus having some beneficial economic effects.   Similarly, increased local procurement of meat, dairy products, or fresh fruit and vegetables, due to the influx of new population, may encourage more intensive grazing and truck farming, with possible resulting beneficial or detrimental changes in land use patterns.   Or this unplanned sudden population increase may tax the capacity of local indoor or outdoor recreational facilities beyond design limits, sometimes to the detriment or destruction of these resources, or force the conversion of wild lands to more developed recreational areas.

Some activities can affect the present or potential suitability of land for certain uses, rather than its availability.   For example, the establishment of an industrial complex near a housing area would seriously limit the adjacent land for use as a school site or for additional housing.   On the other hand, where the adjacent land is being used for heavy industry, sanitary landfill, or warehouse, its potential would be much less affected.   Thus, the ramifications of the proposed project may reach far beyond the perimeter of the project area in diverse ways.

Ecology

Ecology is the study of the interrelationships between organisms and their environment.   Based upon this definition, *all* the subject areas discussed in this section would constitute a part of the overall category of ecology.   In the context of this discussion, however, the category is utilized to include those considerations covering living animals and plant species.

Interest in plant and animal species, especially those becoming less common, prompted the beginnings of modern environmental concern in the mid 1950's.   The general recognition that man was seriously disturbing organisms in the ecosystem without intending to do so, caused ripples of concern, disbelief, and protest which are still with us.   While it has always been recognized that many species have been crowded out of their habitats, and that others have been deliberately exterminated, the gradual comprehension of the fact that man was killing many species, especially through use of broadspectrum pesticides, came as a distinct shock to the scientific community.   Even greater public controversy was generated by groups which actively pressured governmental agencies for enactment of legislation to prevent recurrence of such widespread detrimental impacts.   Modern legislation requiring assessment of likely effects, before initiation of a project, is an outgrowth of these movements of the 1960's.

It is generally agreed that an aesthetically agreeable environment includes as many species of native plants and animals as possible.   In many ways, one may measure the degradation of environments by noting the decrease in these common wildlife species. Since many types of outdoor activities are based directly on wildlife species, there may be economic, as well as moral and aesthetic bases, for maintaining large, healthy populations. The values derived from hunting and fishing activities are the difference between existence and relative affluence for many persons engaged in services connected with these outdoor recreational pursuits.

In considering the impact of man's activities on the biota, it can be determined that there are at least three separable types of interests. The first, *species diversity*, includes the examination of all types and numbers of plants and animals considered as species, whether or not they have been determined to have economic importance or any other special values. The second general area, *system stability*, is basically concerned with the dynamics of relationships among the various organisms within a community.

A third important area, *wildlife management*, deals with species known to have some recreational and economic value, and are usually managed by state or other conservation departments.

All the areas in ecology are very difficult to quantify, often being almost impossible to present in terms familiar to scientists of other disciplines. Furthermore, there are literally millions of possible pathways in which interactions among the plants, animals, and environment may proceed. To date, scientists knowledgeable in the field have been able to trace and analyze only a small minority of these, although thousands more may be inferred from existing data. Thus, many of the impacts predicted cannot be absolutely verified. Other interactions are probably correct by comparison with known cases involving similar situations, while many more are simply predicted on the basis of knowledge and experience in a broad range of analogous, although not closely similar, systems.

The question of chance effects is also an important one in ecology. One may be able to say that the likelihood of serious impact following a certain activity is low, based on available experience. This is definitely *not* the same as saying that the impact, if it develops, is not serious. The impact may be catastrophic, at least on a regional basis, once it develops. When one works with living organisms, the possibility of spread from an area where little chance of damage exists to one in which a greater opportunity for harm is present is itself a very real danger. The vectors of such movement cannot be predicted with any accuracy; however, the basic principles best kept in mind are simple enough. Any decrease in species diversity tends to also decrease the stability of the ecosystem, and any decrease in stability increases the danger of fluctuations in populations of economically important species.

Many other scientific disciplines are often closely related to biology. When the question of turbidity of water in a stream is examined, for example, it will be found that this effect is not only displeasing to the human observer, but has ecological consequences also. The excessive turbidity may cause eggs of many species of fish to fail to develop normally. It may even, in extreme cases, render the water unsuitable for the very existence of several species of fish. The smaller animals and the plant life once characteristic of that watershed may also disappear. Thus, the turbidity of the water, possibly caused by landclearing operations in the stream watershed, may have effects ramifying far beyond the original, observed ones. Similarly, almost all effects which are observed relating to the quality of water will also have some ecological implications, in addition to those already of interest from a water supply viewpoint.

Since it was the observation of damage to the biological environment that was the cause of most of the interest in ecology in the past decade, we must recognize that there is almost no activity which takes place that does not have some ecological implications. These may be simply aesthetic in nature, damaging the appearance of a favorite view, for example. They may also be symptoms of effects which could be harmful to man if left alone for years to come, such as pesticide accumulation by birds and fish. If we are to view the area of biology, or ecology, in perspective, we must realize that it includes a wide variety of messages to man. These should be interpreted as skillfully as possible, if man's future is to be assured.

Sound

Noise is one of the most pervasive environmental problems. The Report to the President and Congress on Noise indicates that between 80–100 million people are bothered by environmental noise on a daily basis and approximately 40 million are adversely affected in terms of health.[3] Relative to the occupational environment, hearing loss primarily due to noise is considered to be the leading occupational disability.[4]

Since noise is a by-product of human activity, the area of exposure increases as a function of population growth, mobility, and such activities as power generation.

A variety of sources produce noise potentially hazardous to hearing, depending upon the intensity and duration of exposure. These include transportation systems, construction equipment, industrial activities, and many common appliances. In addition, speech communication interference problems are common in the environment, and community annoyance is increasingly precipitated by noise generating activities, particularly in populated areas.

The health effects of noise are substantial. It has been reported that 50 to 70 percent of the United States population is annoyed by noise on a daily basis[3]; resulting social and psychological stresses are of major concern to the scientists and planners. The implicated health related effects due to noise include:

1. Permanent or temporary hearing loss;
2. Sleep interference;
3. Increased human annoyance;
4. Communication interference resulting in reduced worker efficiency;
5. Impairment of mental and creative types of work performance; and
6. Possible increase in usage of drugs like sleeping pills, as a method of adaptation to noise stress.[5]

Figure 4.5 summarizes these and other impacts on human activity.

Damage to physical objects is another important consideration. Many natural and man-made features in the environment have become increasingly vulnerable to an ever expanding technology, of which noise is a by-product. Damages associated with noise exposure include:

1. Structural impairment;
2. Property devaluation; and
3. Land use incompatability.

This concern may be supported by considering the damages which are currently being sought by various plaintiffs for transportation noise, amounting to nearly $4 billion.[6]

It has already been noted how noise may affect human health and land use integrity. If a noise has an adverse impact on human physical and mental health, it is likely that the ecosystem (specifically animal life in an exposed area) is also being affected. Chronic noise annoyance and distraction may lead to (1) human error in handling and disposal of hazardous materials,

Physiological Effects
1. Vasoconstriction
2. Gastrointestinal modification
3. Endocrine stimulation
4. Respiratory modification
5. Galvanic skin resistance alteration

Hearing Impairment
1. Permanent/temporary hearing loss
2. Recruitment
3. Tinnitus

Communication Interference
1. Aural-face-to-face; telephone
2. Visual-distortion; color blindness

Task Interference
1. Reduced production
2. Increased error rate
3. Extended output

Sleep Interference
1. Electroencephalographic modification (EEG)
2. Sleep stage alteration
3. Awakening
4. Medication

Personal Behavior
1. Annoyance
2. Anxiety-nervousness
3. Fear
4. Misfeasance

Fig. 4.5 Human activity impacts resulting from increased noise stress.

thereby potentially impacting land, air, and water quality, as well as (2) disrupting harmonious social interaction by creating minor upheavals and disagreements.

On the other hand, because noise restricts the scope of land use, it also tends to depreciate the value of impacted property, including undeveloped as well as developed land. Therefore, the impact of noise may be far reaching, having a potentially significant impact on nearly every other environmental area.

Human Aspects

Men everywhere react to situations as they define them, and if men define a situation as real, then that situation is real in its consequences. This tendency has become a principle of advertising, public and community relations, and "image management." The fact that scientists and engineers think a solution of their own requirements is perfectly rational, economic, and altogether good, may be beside the point. If that solution provokes a public controversy because numerous people and organizations believe it threatens a certain quality of life which they value, then the consequences will be real. Hence, there is the great practical importance of sociopsychological thinking by environment-conscious planners and managers.

Environment is surroundings. Social environment is people surroundings: human beings and their products, their property, their groups, their influence, their heritage. Such are the surroundings of almost any undertaking. There is no one social environment; there are many. Each event—be it the construction of a major facility, a reservoir or power project, proposed legislation, etc., as long as it is at a different place—has its own social environment, its own surroundings.

The effects of a project or plan on people and people's responses may be direct and immediate or remote and attenuated. But it is likely that people are somehow, sooner or later, implicated. And this is apt to be the case even if an activity occurs on a deserted island, miles from human habitation, and the action is triggered by electronic push buttons.

Prerequsite to any rational assessment of human impacts and responses is an inventory and depiction of the relevant social environment. It applies equally well to a wide variety of event-environment situations, and some straightforward observation and fact gathering is all that is necessary.

First, the location of the event itself is established. This can be done on a map having lines and boundaries that have been established by law (town, city, county, state). Location can also be described in terms of topography and physical dimensions; near a river, on a hill, two miles from a freeway, etc. Both means of placing the event may be necessary.

A place (with its people) may be a community or a neighborhood; on the other hand, it may be only a settlement, housing people who have so little in common that they constitute neither a neighborhood nor a community. It is important to learn just what kind of place, socially and politically, one is dealing with. To this end, more questions must be investigated.

Having located the place, the next question is: What are the place's resources upon which people have become dependent? What are the hopes and prospects which they hold dear? This part of environmental description calls for some of the same knowledge that is generated by those who analyze biological and physical environments—the conditions and the resources of the earth, water, air, and climate. The student of social environment, however, is only concerned with these things to the extent that people have come to value them, use them, and require them. This extent and its consequences may both be considerable. People are inclined to fear that their way of life will be damaged or disrupted if the resource base is altered. Their fear is quite understandable.

People, place, and resources, each element acting on the others, produce land uses. A land use is literally the activity and the purpose which a piece of land—a lot, an acreage, an acre—has been put to by people. Uses are mapped and analyzed by many environmental scientists, businessmen, and public officials. Patterns and changes in land use are identified as a basis for locating stores, highways, utilities, and schools. Millions of dollars (or political fortunes) can be lost or made, whether as profits or tax revenues, on accurate predictions of land use trends—from agricultural to residential, or from industrial to unused, for example.

Like many things in society, land uses are never completely stable, and they may change very rapidly. It all depends on what is happening to the *people*—their numbers, their characteristics, their distribution—and to their economy and technology. Therefore, the person assessing environmental impacts, who wants to predict outcomes and weigh alternatives, must know the land use patterns and population trends of one or more places. At the same time, he must figure the economic dimension of the social environment. (In this connection, note the attributes classified as economics.)

So far in this brief account, only what teachers and research scientists call "human ecology" has been introduced. But that is only half of the social environment. Project managers must also assess the political realities of the place in which they would locate their projects and their activities. For engineers, especially, this seems to be difficult. They are used to thinking and working with physical things and with tools from the physical sciences. Social considerations are not their forte. Nevertheless, engineering managers and decision makers today, as never before, perhaps must reckon with human stubbornness and controversy. This is to say, they must anticipate and calculate the political reactions which their work is bound to produce. And they will engage in social engineering insofar as they act upon these considerations.

Because the essential ingredient of politics is power, and power is generated in organizations of people, the wise planner/manager will ask, "What are the organizations in this place, or with a stake in it, that I must reckon with?" State and local governments, business corporations, property owners associations, environmental groups, families; these are some of the kinds of organizations that may be present. How big are they? How powerful and influential are they? How is their policy making done—by what persons and what procedures? Have they enacted laws or regulations that could or should affect major projects? Local and state government land use plans, zoning regulations, and building codes, are examples.

An organization may react favorably, unfavorably, or neutrally. The position an organization takes, as well as its capacity to generate broader support for its policy and to execute it successfully, will depend upon whether and how its members and its public believe their quality of life will be affected by the proposed new project.

Finally, community needs (the overall effects on the local community and public facilities operation) change with changes in the population, human resources, and community facilities. As such, these needs deal with potential effects on local housing, schools, hospitals, and local government operations.

Economics

Measurement of economic impact may be as simple as estimating the change in income in an area, or as complicated as determining the change in the underlying economic structure and distribution of income. Generally, effects may be examined for impact on conditions (income, employment) or structure (output by sector, employment by sector). These effects may be measured as impacts on the stock of certain resources or the flow of an economic parameter. We will discuss briefly the value of assets (stock), employment, income, and output as categories of variables.

Community or regional assets may be affected by project activity and these assets may or may not be irreplaceable. The change in value of land and natural resources is an indicator of change in the stock or quantity of certain resources—for example, minerals—which are used in the conduct of social and economic activities. The category of land and natural resources which are not readily replenished by additional economic activities includes coal, a natural resource which, once mined and utilized, cannot be replaced.

This category of economic change is important to decision makers because the extent to which the quantity of irreplaceable resources is changed will become increasingly more controversial as real or feared shortages of these resources develop.

The value of structures, equipment, and inventory is an indicator of change in the stock or quantity of resources such as buildings, trucks or furniture which are used in the conduct of man's social and economic activities. This group of resources represents capital stocks that are replaceable by additional economic activity. For example, it would be possible to reconstruct a building elsewhere if it were rendered useless by project activity. If a project were to make some vehicles obsolete, replacement with other, newer alternatives would be possible.

Total employment effects relate to all full-time and part-time employees in a region, on the payroll of operating establishments, or other forms of organization, who worked or received pay for any part of a specified period. Included are persons on paid sick leave, paid holidays, and paid vacations during the pay period. Officers of corporations are included as employees. Total employment can be affected by direct demand for services to perform a specified task or by indirect demand and secondary and tertiary activities that affect the requirements for goods and services.

Total income for a region refers to the money income of people employed in the conduct of economic activities in the region. This income normally comes from salaries and wages paid to the individuals in return for services performed. Included are incomes from social security, retirement, public assistance, welfare, interests, dividends, and net income from property rental.

Incomes are most easily affected by changes in purchasing patterns in the region. The magnitude of a project's potential effect is related to the extent to which purchases of goods and services in the region are significant and will increase or decrease.

Output can be defined as goods and services produced by sectors of the economy in the region. Indicators of regional output are: (1) value added to a product as a result of a manufacturing process, (2) gross receipts for service industries, (3) total sales from the trade sector, and (4) values of shipments. Output can be affected by direct and indirect expenditure and employment changes.

Other areas of potential impact relate to income distribution, the distribution of production by sector, governmental expenditures, and revenue collections by governmental units. The possible impact categories are extensive, and this brief introduction touches in a few of the more widely recognized areas.

Resources

The United States entered a new era of its history in the early 1970's. Shortages of many commonplace items, such as meat, building materials, and gasoline, fluctuated from adequacy to virtual nonavailability in many sections of the country. The period beginning in the early 1970's has been termed "the era of shortages" by many commentators surveying the American scene.

The rampant gas and oil shortage came as no surprise to experts in the economic and energy fields, but for the first time, the American public became aware that the question of energy supply could dramatically affect the quality of day-to-day life. Federal agencies experienced cutbacks in allocations for fuel and petroleum products. Interest in energy conservation was stimulated as a result of these shortages; magazine articles, news broadcasts, and newspapers pointed out energy conservation methods, presented information on energy supply, and exposed many groups involved with wasteful practices.

The energy situation was not the only concern resulting from the shortages experienced in 1973 and 1974. Increasing realization of the fact that many of our domestic mineral resources are rapidly approaching depletion, as shown in Table 4.5, has prompted renewed interest in the search for new materials which could be substituted for heavily impacted resources. In addition, the question of obtaining raw materials has generated concern. The United States' increasing dependence on foreign sources for petroleum, minerals, and other non-renewable critical resources, along with concern for the balance of payments and national security, has increased interest in conserving and recycling resources, and has renewed the search for alternate sources of energy.

Environmental quality is directly linked with the use and procurement of energy. The continued degradation of air and water resources, the irrevocable loss of wilderness areas, and land use planning dilemmas are problems which must be dealt with in the development of resources. Environmental considerations have delayed the construction of the Alaskan pipeline and offshore drilling projects, and promise to be a major obstacle in the development of the western coal fields. Air pollution resulting from emissions from the combustion of fossil fuels in engines and furnaces is also another cause for concern. The necessity of providing a safe and healthful environment is another motive for the development of alternate energy sources which are also non-polluting.

Another environmental attribute which may be thought of as a resource is aesthetics. Although difficult to measure or quantify, the environment, as apprehended through hearing, sound, sight, smell, and touch, is important to everyone, although each individual perceives this environment and re-

TABLE 4.5   Proved World Reserves of Selected Minerals

More than 100 years	Columbium
	Potash
	Phosphorus
	Magnesium
51–100 years	Iron ore
	Chromite
	Nickel
	Vanadium
	Cobalt
	Asbestos
	Molybdenum
26–50 years	Manganese
	Bauxite
	Platinum
	Titanium
	Antimony
	Sulfur
15–25 years	Copper
	Lead
	Tin
	Zinc
	Tungsten
	Barite
10–15 years	Mercury
	Silver

SOURCE: "Critical Imported Materials," Council on International Economic Policy, USGPO, Washington, D.C., December 1974.

sponds to it differently.   Project planners today are faced with increased pressures to incorporate not only functional engineering and cost aspects, but also to include aesthetic considerations in every planning activity.

4.3   DETERMINING ENVIRONMENTAL IMPACT

The distinction between environmental impact and changes in environmental attributes is that changes in the attributes provide an indication of changes in the environment.   In a sense, the set of attributes must provide a model for the prediction of all impacts.   The steps in determining environmental impact are:

1.   Identification of impacts on attributes;

2. Measurement of impacts on attributes;
3. Aggregation of impacts on attributes to reflect impact on the environment.

With and without the Project

The conditions for estimating environmental impact are measurement of attributes with and without the project or activity under consideration at a given point in time. Fig. 4.6 indicates the measure of an attribute with and without an activity over time. From this example, it can be seen that the measure of the attribute may change over time, without the activity. Therefore, the impact must be measured in terms of the "net" change in the attribute at a given point in time.

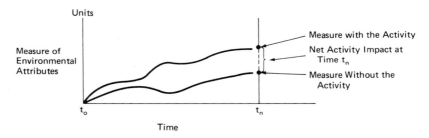

Fig. 4.6   Measure of impact with and without activity.

This concept of impact is used to avoid problems of comparing the present measure (without the activity) with the future measure (with the activity). The difficulty is that data for a "with activity" and "without activity" projection of impacts are difficult to obtain, and results are difficult to verify. However, several well established forecasting techniques are available for establishing the "without" project condition, based on assumptions made for alternative futures. Quantifiable attributes, especially, can be forecast using past data and mathematical trend forecasting techniques.[7]

Identifying Impacts

The list of environmental attributes that might be evaluated is practically infinite because *any* characteristic of the environment is an attribute. Therefore, it is necessary to reduce the number of attributes to be examined. Thus, duplicative, redundant, difficult to measure, and obscure attributes may be eliminated in favor of those that are more tractable. This procedure is valid only if the remaining attributes reflect all aspects of the environment. This means that some attributes, difficult to measure or conceptualize, may remain to be dealt with.

Thus, identification of impacts is based on a review of potentially impacted attributes to determine whether they will be affected by the subject activity.

Characteristics of the Base

Conditions Prior to the Activity.   The nature of the impact is determined by the conditions of the environment prior to the activity.   Base data are information regarding what the measure of the attributes would be (or is) prior to the activity at the project location.   Because the measurement and analysis of environmental impact cannot take place without base data, identifying the characteristics of the base is critical.

Geographic Characteristics.   There may be significant differences in impact on attributes for a given activity in different areas. Geographical location is, therefore, one of the factors that affects the merit or relative importance of considering a particular attribute.   For example, the impact of similar projects on water quality in an area with abundant water supplies versus impact in an area with scarce water resources would differ significantly.   The spatial dispersion of different activities introduces one of the difficult elements in comparing one activity and its impact with another.

Temporal Characteristics.   Time may also pose problems for the impact analysis.   It is essential to ensure that all impacts are examined over the same projection time period.   Furthermore, to adequately compare (or combine) activity impacts, it is necessary that the same time period (or periods) apply.

Role of the Attributes

Although potential effects of impacts can be considered as effects on definite discrete attributes of the environment, the impression must not be created that actual impacts are correspondingly well categorized.   That is, nature does not necessarily respect man's discrete categories. Rather, actual impacts may be "smears" comprised of effects of varying severity on a variety of interrelated attributes.   Many of these interrelationships may be handled by noting the attributes primarily affected by activities and by utilizing the descriptions contained in the descriptor package in Appendix B to point out the secondary, tertiary, etc., effects.

Measurement of Impact

Identifying the impact of a project on an attribute leads directly to the second step of measuring the impact. Ideally, all impacts should be translatable into common units. This is, however, not possible because of the difficulty in defining impacts in common units (e.g., on income and on rare and endangered species). In addition to the difficulties in quantitatively identifying impacts, are the problems that arise because quantification of some impacts may be beyond the state of the art. Thus, the problems of measuring and comparing them with quantitative impacts are introduced.

Quantitative Measurements

Quantitative measurements of impact are measures of projected change in the relevant attributes. These measurement units must be based on a technique for projecting the changes into the future. The changes must be projected on the basis of a no-activity alternative. One difficulty in assessing the quantitative change arises from the fact that changes in attributes may not be in common units. In addition, there are difficulties in assessing the changes in the attributes through the use of projection techniques.

Qualitative Measurements

Changes in some attributes of the environment are not amenable to measurement. The attribute may not be defined well enough in its relationship to overall environment to determine what the most adequate measurable parameter might be. Therefore, instead of a specific measure, a general title and definition may be all that is available. In such cases, it may be necessary to rely on expert judgment to answer the question of how attributes will be affected by the subject project.

Comparison Between Attributes

In the development of any technique or methodology for environmental impact analysis, inevitably a time will come when someone asks the question "How do you compare all these environmental parameters with one another?" And, as is usually the case, long-lasting and frequently heated arguments follow, with the final result generally being the consensus that there is no conclusion. Indeed, the question of comparing "apples and alligators" or even worse, "Biochemical Oxygen Demand" and "Public Sector Revenue" bears no simple or well-defined solution. There have been some attempts at developing schemes for making numerical comparisons, which will be discussed in more detail in Chapter 5.

Another interesting procedure for developing such information is also available—a modification of the Delphi technique.[2]

The Delphi technique is a procedure, developed at the Rand Corporation, for eliciting and processing the opinions of a group of experts knowledgeable in the various areas involved. A systematic and controlled process of queing and aggregating the judgments of group members is used, and stress is placed upon iteration with feedback to arrive at a convergent consensus. The weighing system discussed in the following section does not include all the elements of a Delphi technique. In addition, results of these ranking sessions need further study, feedback, and substantive input from field data.

The weighing procedure can be accomplished in a very simple manner. A deck of cards is given to each person participating in the weighing. Each card names a different technical specialty. Each of the participants is then requested to rank the technical specialties according to their relative importance to explain changes in the environment from major activities. Then each individual is asked to go back through the list, making a pairwise comparison between technical specialties, beginning with the most important one. The most important technical specialty is compared with the next most important by each individual, and the second technical specialty is assigned a percentage. This assignment is to reflect the percentage of importance of the second technical specialty with respect to the first. For example, the first technical area would receive a weight of 100 percent, and the second most important technical area might be considered by the specialist to be only 90 percent as important as the first. Then the second and third most important technical specialties are compared, and the third most important area is assigned a number (for example, 95 percent) as its relative importance compared to the second most important technical specialty. A sample diagram of the comparison is presented in Fig. 4.7.

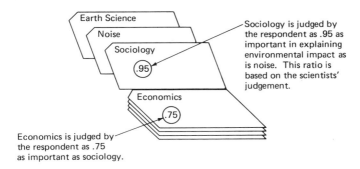

Fig. 4.7 **Pairwise comparison of environmental attributes.**

The formula for weighing the technical specialties is:

$$W_{ij} = \frac{V_{ij}}{\sum\limits_{i-1}^{n} V_{ij}} P \qquad (i = 1, 2, 3, \ldots, n)$$

$$V_{ij} = \begin{cases} 1 & (i = 1) \\ V_{i-1j} X_{ij} & (i = 2, 3, \ldots, n) \end{cases}$$

where

W_{ij} = weight for the ith technical specialty area by the jth scientist;

n = number of technical specialties;

P = 1000: total number of points to be distributed among the technical specialties;

X_{ij} = the jth scientist's assessment of the ratio of importance of the ith technical specialty in relation to the $(i - 1)^{th}$ technical specialty;

V_{ij} = measure of relative weight for the ith technical specialty area by the jth scientist.

To accomplish the second part of this technique, i.e., to rank attributes within a technical specialty, each scientist independently ranks attributes in his own specialty.

The information from these pairwise comparisons then can be used to calculate the relative importance of each of these technical specialty areas; a fixed number of points (e.g., 1000) is distributed among the technical specialties, according to individual relative importance.

After the weights are calculated from one round of this procedure, the information about the relative weights is presented again to the experts, a discussion of the weights is undertaken, and a second round of pairwise comparisons is made. The process is repeated until the results become relatively stable in successive rounds.

In a demonstration of this method, an interdisciplinary group of college graduates with very little interdisciplinary training was asked to rate the following areas according to their relative importance in environmental impact analysis, and to distribute a 1000-point total among the categories:

1. Air quality
2. Ecology
3. Water quality
4. Aesthetics
5. Economics
6. Transportation

TABLE 4.6 Results of Modified Delphi Procedure for Comparing Environmental Areas

	Before Interdisciplinary Study		After Interdisciplinary Study
Area	Average Point Distribution	Area	Average Point Distribution
Water	125	Water	128
Air	122	Air	126
Natural resources	109	Natural resources	105
Health	100	Ecology	93
Ecology	97	Health	88
Land use	81	Earth science	87
Earth science	79	Land use	78
Economics	62	Sociology	64
Sociology	60	Noise	62
Transportation	56	Economics	62
Aesthetics	54	Transportation	61
Noise	53	Aesthetics	46
Total	1000	Total	1000

7. Earth science
8. Sociology
9. Natural resources and energy
10. Health science
11. Land use
12. Noise.

After a thorough group study of all 12 areas, the group was asked to rate the areas again. The results, shown in Table 4.6, indicate that although some relative priorities changed, the points allocated to each category remained essentially the same. Similar ratings may be developed for attributes within each group.

It should be emphasized that this procedure as described is considered only a tool for arriving at group decisions. Different groups would probably arrive at different decisions, and any application directed toward comparison between attributes should be made in the context of a specific planning situation.

Aggregation

After measuring project impacts on various attributes, two aggregation problems must be addressed. The first problem deals with how to aggregate among the different attributes (quantitative *and* qualitative) to arrive at a single measure for activity impact. Doing this would involve expressing the various impact measures in common units. Then, a method for aggregating the impacts on a specific attribute must be identified. (Some methodologies utilize a weighting procedure to accomplish this.) Finally, the impacts may be summed and compared with the impact of an alternative activity. A graphical method for accomplishing this summary is discussed in Appendix C.

Secondary Impacts

Secondary or indirect consequences for the environment should be addressed, especially as related to infrastructure investments that "stimulate or induce secondary effects in the form of associated investments and changed patterns of social and economic activity." These effects may be produced "through their impact on existing community facilities and activities, through induced new facilities and activities, or through changes in natural conditions." A specific example calls out possible changes in population patterns and growth that may have secondary and indirect effects upon the resource base, including land use, water, and public services. In the biophysical environment, the secondary impacts can also be important.

To illustrate the nature of interrelationship between environmental attributes, consider, as an example, an activity which involves extensive removal of vegetation in a water shed. The environmental attribute which would be indicated as being affected by this activity would be erosion. The examination of this attribute would lead to other potentially impacted attributes such as dissolved oxygen, suspended solids, nutrient concentration (which may stimulate algae growth) and cause a change in community maintenance (the numbers of organisms and consumption of aquatic species in the stream). The pH of the stream could be affected by the growth of algae, and this, in turn, could effect the concentration of many of the chemicals in the stream by changing their solubility. Change in each of the chemical constituents affected could trigger further change in the complex system. Excessive growth of algae could, at some location, result in high BOD values and loss of oxygen from the stream. Clearly, the interrelationships would not be limited to the stream, for evolution of gases from decomposition could create air pollution problems. This and/or the green color of the stream could affect land use and cause adverse social and economic effects.

Cumulative Impacts

A single activity may produce a negligible effect on the environment. However, a series of similar activities may produce cumulative effects on certain aspects of the environment. This raises the question of how to deal with these potential cumulative effects. The most obvious solution is to prepare impact assessments on broad programs rather than on a series of component actions. Unfortunately, the definition of activities at the program level may be so vague as to preclude identification of impacts on the attributes of the environment. Nevertheless, review of activities at the program level, requiring enough detail to evaluate impacts, is the best way to handle the problem of cumulative impacts.

4.4 REPORTING FINDINGS

Results of the impact analysis process are documented as one of the following:

1. An assessment
2. A negative declaration statement
3. A draft statement
4. A final statement.

It is useful to consider displaying the results in a way that makes it easy to comprehend total impact from a brief review. One suggested method for doing this is by displaying the impacts on the summary sheet included in Appendix C.

Details of the specific format for an environmental impact analysis documentation are given by individual agency guidelines. These guidelines should be consulted and followed for each analysis.

REFERENCES

1. *Environmental Quality*, Third Annual Report of the Council on Environmental Quality, U.S. Government Printing Office (USGPO), August 1972.

2. Jain, R. K. *et al.*, "Environmental Impact for Army Military Programs," *Interim Report D-13/AD 771062*, U.S. Army Construction Engineering Research Laboratory (USACERL), December 1973.

3. U.S. Environmental Protection Agency, "Report to the President and Congress on Noise," December 1971 (USGPO, 1972).

4. Bragdon, Clifford R., *Noise Pollution: The Unquiet Crisis*, University of Pennsylvania Press, Philadelphia, 1971.

5. Bragdon, Clifford R., "Community Noise: A Status Report," paper presented at the 84th meeting of the Acoustical Society of America, Miami, Florida, November 1972.

6. Bragdon, Clifford R., "Noise Control in Urban Planning," *Journal of Urban Planning and Development Division* **99,** *No. 1,* ASCE (1973).

7. "Handbook of Forecasting Techniques," IWR Report 75-7, U.S. Army Engineer Institute for Water Resources, Ft. Belvoir, Virginia, December 1975.

8. Environmental Quality, Fourth Annual Report of the Council on Environmental Quality, USGPO, September 1973.

5

Impact Assessment
Methodologies

Many methodologies have been developed which allow the
user to respond in a substantive manner to CEQ requirements when preparing an EIA/EIS. Presented in this chapter is a review and analysis of some of these environmental
impact assessment methodologies. The purpose of this
discussion of impact assessment methodologies is to (a)
acquaint the reader with the different general types of
methodologies; and (b) provide illustrative examples of
available methodologies in each category. The initial review and analysis of impact analysis methodologies was
first completed by Warner and Preston in 1974[23]. The discussion here draws substantial information from their
work.

5.1 CHOOSING A METHODOLOGY

Depending upon the specific needs of the user and the type
of project being undertaken, one particular methodology

may be more useful than another. Each individual must determine which tools best fit a given task. To select the most appropriate tools, the following key considerations may be useful.

Use.   Is the analysis primarily a decision or an information document? (A decision document is vital for determining the best course of action, while an information document primarily reveals implications of the selected choices.) A decision document analysis generally requires greater emphasis on identification of key issues, quantification, and direct comparison of alternatives.   An information document requires a more comprehensive analysis and concentrates on interpreting the significance of a broader spectrum of possible impacts.

Alternatives.   Are alternatives fundamentally or incrementally different? If differences are fundamental (such as preventing flood damage by levee construction as opposed to flood plain zoning), then impact significance should be measured against some absolute standard, since impacts will differ in type as well as size. On the other hand, incrementally different alternative sets permit direct comparison of impacts and a greater degree of quantification.

Public Involvement.   Does the role of the public in the analysis involve substantive preparation or review? Substantive preparation allows use of more complex techniques, such as computer or statistical analysis that might be difficult to explain to a previously uninvolved but highly concerned public. A substantive preparation role will also allow a greater degree of quantification or weighting of impact significance through the direct incorporation of public values.

Resources.   How much time, skill, money and data, and what computer facilities are available? Generally, more quantitative analyses require more of everything.

Familiarity.   Is the preparer familiar with both the type of action contemplated and the physical site? Greater familiarity will improve the validity of a more subjective analysis of impact significance.

Issue Significance.   How big an issue is being dealt with? All other things being equal, the bigger the issue, the greater the need to be explicit, to quantify, and to identify key issues. Arbitrary weights or formulae for trading off one type of impact (e.g., environmental) against another (e.g., economic) become less appropriate.

Administrative Constraints.   Are choices limited by agency procedural or format requirements?   Specific agency policy or guidelines may rule out some tools by specifying the range of impacts to be addressed, the need for analyzing the trade-offs, or the time frame of analysis.

5.2 CATEGORIZING METHODOLOGIES

The various methodologies examined can be divided into six types, based upon the way impacts are identified.

Ad Hoc.   These methodologies provide minimal guidance for impact assessment beyond suggesting broad areas of possible impacts (e.g., impacts upon flora and fauna, impacts on lakes, forests), rather than defining specific parameters to be investigated.

Overlays.   These methodologies rely upon a set of maps of a project area's environmental characteristics (physical, social, ecological, aesthetic). These maps are overlaid to produce a composite characterization of the regional environment.   Impacts are identified by noting the impacted environmental characteristics within the project boundaries.

Checklists.   These methodologies present a specific list of environmental parameters to be investigated for possible impacts; they do not require establishing direct cause-effect links to project activities. They may or may not include guidelines about how parameter data are to be measured and interpreted.

Matrices.   These methodologies incorporate a list of project activities with a checklist of potentially impacted environmental characteristics. The two lists are related in a matrix which identifies cause-effect relationships between specific activities and impacts.   Matrix methodologies may either specify which actions impact which environmental characteristics, or may simply list the range of possible actions and characteristics in an open matrix to be completed by the analyst.

Networks.   These methodologies work from a list of project activities to establish cause-condition-effect relationships. They are an attempt to recognize that a series of impacts may be triggered by a project action. Their approaches generally define a set of possible networks and allow the user to identify impacts by selecting and tracing out the appropriate project actions.

Combination Computer-Aided. These methodologies use a combination of matrices, networks, analytical models, and a computer-aided systematic approach to (a) identify activities associated with implementing major federal programs, (b) identify potential environmental impacts at different user levels, (c) provide guidance for abatement and mitigation techniques, (d) provide analytical models to establish cause-effect relationships to quantitatively determine potential environmental impacts, and (e) provide a methodology and a procedure to utilize this comprehensive information in responding to requirements of EIS preparation.

5.3 REVIEW CRITERIA

To serve the purposes of NEPA, an environmental impact assessment must effectively deal with four key problems: (a) impact identification, (b) impact measurement, (c) impact interpretation, and (d) impact communication to information users.

Experience with impact assessments to date has shown that a set of methodology evaluation criteria can be defined for each of these four key problems. This review criteria can be used for analyzing a methodology and determining its weaknesses and strengths. The criteria are as follows:

Impact Identification.

Comprehensiveness. A full range of impacts should be addressed, including: ecological, physical-chemical pollution, social-cultural, aesthetic, resource supplies, induced growth, induced population or wealth redistributions, and induced energy or land-use patterns.

Specificity. A methodology should identify specific parameters (subcategories of impact types, i.e., detailed parameters under the major environmental categories of air, water, ecology, etc.) to be examined.

Isolating Project Impacts. Methods to identify project impacts, as distinct from future environmental changes produced by other causes, should be required and suggested.

Timing and Duration. Methods to identify the timing (short-term operational versus long-term operational phases) and duration of impacts should be required. (Data sources should also be listed for impact measurement and interpretation.)

Data Sources. Identification of the data sources used to identify impacts should be required. (Data sources should also be listed for impact measurement and interpretation.)

Impact Measurement.

Explicit Indicators.   Specific measurable indicators to be used for quantifying impacts upon parameters should be suggested.

Magnitude.   A methodology should require and provide for the measurement of impact magnitude, as distinct from impact significance.

Objectivity.   Objective rather than subjective impact measurements should be emphasized.

Impact Interpretation.

Significance.   Explicit assessment of the significance of measured impacts on a local, regional and national scale should be required.

Explicit Criteria.   A statement of the criteria and assumptions employed to determine impact significance should be required.

Uncertainty.   An assessment of the uncertainty or degree of confidence in impact significance should be required.

Risk.   Identification of any impacts having low probability but high damage or loss potential should be required.

Alternatives Comparison.   A specific method for comparing alternatives, including the "no project" alternative, should be provided.

Aggregation.   A methodology may provide a mechanism for aggregating impacts into a net total or composite estimate.   If aggregation is included, specific weighting criteria or processes to be used should be identified.   The appropriate degree of aggregation is a hotly debated issue on which no judgment can be made at this time.

Public Involvement.   A methodology should require and suggest a mechanism for public involvement in the interpretation of impact significance.

Impact Communication.

Affected Parties.   A mechanism for linking impacts to the specific affected geographical areas or social groups should be required and suggested.

Setting Description.   A methodology should require that the project setting be described to aid statement users in developing an adequate overall perspective.

Summary Format.   A format for presenting, in summary form, the results of the analysis should be provided.

Key Issues.   A format for highlighting key issues and impacts identified in the analysis should be provided.

NEPA Compliance.   Guidelines for summarizing results in terms of the specific points required by NEPA and subsequent CEQ guidelines should be provided.

In addition to the above "content" criteria, methodological tools should be evaluated in terms of their resource requirements, replicability, and flexibility. The following considerations, used in arriving at the generalized ratings for these characteristics (shown in Table 5-1) may be useful when considering the appropriateness of other tools. Table 5-1 provides a framework for methodology evaluation.

Resource Requirements.

Data Requirements.   Does the methodology require data that are presently available at low retrieval costs?

Manpower Requirements.   What special skills are required?

Time.   How much time is required to learn to use and/or apply the methodology?

Costs.   How do costs using a methodology compare to costs of using other tools?

Technologies.   Are any specific technologies (e.g., computerization) required to use a methodology?

Replicability.

Ambiguity.   What is the relative degree of ambiguity in the methodology?

Analyst Bias.   To what degree will different impact analysts using the methodology tend to produce widely different results?

Flexibility.

Scale Flexibility.   How applicable is the methodology to projects of widely different scale?

Range.   For how broad a range of project or impact types is the methodology useful in its present form?

Adaptability.   How readily can the methodology be modified to fit project situations other than those for which it was designed?

Comparison of Methodologies.   Methodologies may be rated for their degree of compliance with the 20 content criteria discussed above. Three rating characteristics are suggested as follows:

S = Substantial compliance, low resource needs, or few replicability-flexibility limitations

P = Partial compliance, moderate resource needs, or major limitations

N = No compliance or minimal compliance, high resource needs, or major limitations.

TABLE 5.1 Frame Work for Methodology Evaluation.

	Type	Comprehensiveness	Specificity	Isolate Project Impacts	Timing and Duration	Data Sources	Explicit Indicators	Magnitude	Objectivity	Significance	Explicit Criteria	Uncertainty	Risk	Alternatives Comparison	Aggregation	Public Involvement	Affected Parties	Setting Description	Summary Format	Key Issues	NEPA Compliance	Resource Requirements	Replicability	Flexibility
Adkins	C																							
Dee (1972)	C																							
Dee (1973)	C-M																							
Georgia	C																							
Jain/Urban (73)	CO																							
Jain (74)	M																							
Krauskopf	O																							
Leopold	M																							
Little	C																							
McHarg	O																							
Moore	M																							
New York	M																							
Smith	C																							
Sorensen	NW																							
Stover	C																							
Task Force	C																							
Tulsa	C																							
Walton	C																							
WSCC	A																							

Key to types:
A = ad hoc
O = overlay
C = checklist
M = matrix
NW = network
CO = combination computer-aided

Key to evaluation symbols:
S = substantial compliance, low resource needs, or few replicability-flexibility limitations
P = partial compliance, moderate resource needs, or moderate limitations
N = no or minimal compliance, high resource needs, or major limitations
– = aggregation not attempted

These ratings may be applied to various methodologies in order to choose one best-suited for a particular application. Table 5.1, a summary of methodology evaluation, can be completed as a practical exercise for the methodologies discussed herein or for other emerging methodologies.

5.4 METHODOLOGY DESCRIPTIONS

Nineteen methodologies or tools were examined in detail. The brief description given for each methodology discusses some or all of the following points:

> Methodology type
> General approach used
> Range of actions or project types for which the methodology may be applicable
> Comprehensiveness of the methodology in terms of the range of impacts addressed
> Resources required (data, manpower, time, etc.)
> Limitations of the methodology (replicability, ambiguity, flexibility)
> Key ideas or particularly useful concepts
> Other major strengths and weaknesses as identified by the review criteria.

Because of the brevity and subjectivity of these characterizations, they should not be considered as adequate critiques of the tools examined. They may instead serve as a useful introduction to the range of techniques now evolving. Many other methodologies, beyond those discussed here, are available for use by different agencies. The list of methodologies discussed here should not be considered exhaustive because of the dynamic nature of this subject area.

5.5 METHODOLOGY REVIEW

Interim Report: Social, Economic, and Environmental Factors in Highway Decision Making[1] *[Checklist]*. This methodology is a checklist which uses a +5 to −5 rating system for evaluating impacts. The approach was developed to deal specifically with the evaluation of highway-route alternatives. Because the bulk of parameters used relates directly to highway transportation, the approach may not be readily adaptable to other project types.

The parameters used are broken down into categories of transportation, environmental, sociological, and economic impacts. Environmental parameters are generally deficient in ecological considerations. Social parameters emphasize community facilities and services.

Route alternatives are scored +5 to −5 in comparison with the present state of the project area, not the expected future state without the project.

Since the approach uses only subjective relative estimations of impacts, the data, manpower, and cost requirements are very flexible. Reliance upon subjective ratings without guidelines for such ratings reduces the replicability of analysis and generally limits the valid use of the approach to a case-by-case comparison of alternatives only.

The detailed listing of social and, to a lesser extent, economic parameters may be helpful for identifying and cataloging impacts for other types of projects. An interesting feature of possible value to other analyses using relative rating systems is the practice of summarizing the number and the magnitude of plus and minus ratings for each impact category. The number of pluses and minuses may be a more reliable indicator for alternative comparison, since it is less subject to the arbitrariness of subject weighting. These summaries are additive, and thus implicitly weigh all impacts equally.

Environmental Evaluation System for Water Resources Planning[2] [*Checklist*]. This methodology is a checklist procedure emphasizing quantitative impact assessment. While it was designed for water-resource projects, most parameters used are also appropriate for other types of projects. Seventy-eight specific environmental parameters are defined within the four categories of ecology, environmental pollution, aesthetics, and human interest. The approach does not deal with economic or secondary impacts, and social impacts are partially covered within the human-interest category.

Impacts are measured via specific indicators and formulae defined for each parameter. Parameter measurements are converted to a common base of "environmental quality units" through specified graphs or value functions. Impacts can be aggregated by using a set of pre-assigned weights.

Resource requirements are rather high, particularly data requirements. These requirements may restrict the use of the approach to major project assessments.

The approach emphasizes explicit procedures for impact measurement and evaluation and should therefore produce highly replicable results. Both spatial and temporal aspects of impacts are noted and explicitly weighted in the assessment. Public participation, uncertainty, and risk concepts are not dealt with. An important idea of the approach is the highlighting of key impacts via a "red flag" system.

Planning Methodology for Water Quality Management: Environmental System[3] [*Checklist Matrix*]. This unique methodology of impact assessment defies ready classification, since it contains elements of checklist, matrix, and network approaches. Areas of possible impacts are defined by a hierarchical system of four categories (ecology, physical/chemical, aesthetic, social), 19 components, and 64 parameters. An interaction matrix is presented to indicate which activities associated with water quality treatment projects generally impact which parameters. The range of parameters used is comprehensive, excluding only economic variables.

Impact measurement incorporates two important elements. A set of "ranges" is specified for each parameter to express impact magnitude on a scale from 0 to 1. The ranges assigned to each parameter within a component are then combined by means of an "environmental assessment tree" into a summary environmental impact score for that component. The significance of impacts for each component is quantified by a set of assigned weights. A net impact can be obtained for any alternative by multiplying each component score by its weight factor and summing across components.

The key features of the methodology are its comprehensiveness, its explicitness in defining procedures for impact identification and scoring, and its flexibility in allowing use of best available data.

Sections of the report explain the several uses of the methodology in an overall planning effort and discuss means of public participation. While the data, time, and cost requirements of the methodology when used for impact assessment are moderate, a small amount of training would be required to familiarize users with the techniques.

The methodology possesses only minor ambiguities and should be highly replicable. Because the environmental assessment "trees" are developed specifically for water-treatment facilities, the methodology cannot be readily adapted to other types of projects without reconstructing the "trees," although the parameters could be useful as a simple checklist.

One potentially significant obstacle to use of this approach is the difficulty of explaining the procedures to the public. Regardless of the validity of the "trees," they are devices developed by highly specialized multivariant-analysis techniques, and public acceptance of conclusions reached by their use may be low.

Optimum Pathway Matrix Analysis Approach to the Environmental Decision Making Process: Test Case: Relative Impact of Proposed Highway Alternatives[15] *[Checklist]*. This methodology incorporates a checklist of 56 environmental components. Measurable indicators are specified for each component. The actual values of alternative plan impacts on a component are normalized and expressed as a decimal of the largest impact (on that one component). These normalized values are multiplied by a subjectively determined weighting factor. This factor is the sum of one times a weight for "initial" effects plus ten times a weight for "long-term" effects.

The methodology is used to evaluate highway project alternatives, and the components listed are not suitable for other types of projects. The wide range of impact types analyzed include land use, social, aesthetics, and economic.

The potential lower replicability of the analysis produced by using subjectively determined weighting factors is compensated for by conducting the analysis over a series of iterations and incorporating stochastic error variation in both actual measurements and weights. This procedure provides a basis for testing the significance of differences in total impact scores between alternatives.

The procedure for normalizing or scaling measured impacts to obtain commensurability and testing of significant differences between alternatives are notable features of potential value to other impact analyses and methodologies. These ideas may be useful whenever several project alternatives can be identified and compared.

This methodology may place rather high resource demands, because computerization is necessary to generate random errors and make the large number of repetitive calculations.

Environmental Impact Assessment Study for Army Military Programs[6] *and Computer-Aided Environmental Impact Analysis for Construction Activities: User Manual*[21] *[Combination Computer-Aided].* This is a computer-aided assessment system employing the matrix approach to identify potential environmental impacts. The system relates Army activities from nine functional areas to attributes contained in 11 technical areas of specialty describing the environment. The nine functional areas are: construction; research and development; real-estate acquisition or outleases of land; mission change; procurement; training; administration and support; industrial activities; and operation and maintenance.

Three levels of attributes are identified: detailed level, review level, and controversial attributes. Ramification remarks regarding potential impacts are presented along with mitigation procedures for minimizing adverse impacts. Potential impacts are identified on a need-to-consider scale, using A, B, and C as indicators, instead of a numerical system.

Given the appropriate input information for a particular program, the computer-aided system developed will provide relevant environmental information to allow the user to respond to the requirements of CEQ guidelines. In addition, analytical models are being developed to quantitatively assess the environmental impacts. One such model, the Economic Impact Forecast System (EIFS) is operational at this time.

Significant features of this methodology are: (a) it is cost-effective; (b) it provides analytical models for cause-effect relationships; (c) it is a comprehensive methodology; (d) the output matrix is modified, based upon site-specific input, to produce a project-specific input matrix; (e) it provides information regarding environmental laws and regulations; and (f) it includes information about abatement and mitigation techniques.

This methodology is designed for Army military programs. Its applicability to programs of other agencies is limited and would thus require some systemic modifications.   Problems associated with effective community participation and evaluation of trade-off between short-term areas of environmental resources and long-term productivity are not adequately addressed.

Handbook for Environmental Impact Analysis[7] [*Matrix*].   Employing an open-cell matrix approach, this handbook presents recommended procedures for use by Army personnel in the preparation and processing of environmental impact assessments and statements. The procedures outline an eight-step algorithm in which details of the proposed action and associated alternatives are identified and evaluated for environmental effects in both the biophysical and socioeconomic realms.   Briefly, the procedural steps are outlined as follows:

1. Identify the need for an EIA or an EIS.
2. Establish details of the proposed action.
3. Examine environmental attributes, impact analysis worksheets, and summary sheets.
4. Evaluate impacts, using attribute descriptor package.
5. Summarize impacts on summary sheet.
6. Examine alternatives.
7. Address the eight points of CEQ guidelines.
8. Process final document.

The handbook provides examples of representative Army actions that might have a significant environmental impact (Step 1) and guidance on the identification of Army activities (Steps 2 and 4) for Army functional areas.

Environmental attributes (Steps 3 and 4) are identified and characterized. After evaluating the effect of the proposed action and the alternatives (Step 6) on the interdisciplinary attributes, and summarizing the effects (Step 5), it is recommended that the assessment be documented in the format suggested by the CEQ guidelines (Step 7). Each of the eight points in the CEQ guidelines is discussed in detail, and Army-related examples are presented. In addition, the handbook gives information regarding processing of assessments and statements (Step 8).

Because the methodology is designed for Army military programs, its applicability to programs of other agencies is limited and would require systemic modifications.   In addition, this methodology does not provide the depth and comprehensiveness of environmental information made available by the computer-aided study[6,21] previously discussed.

Evaluation of Environmental Impact Through a Computer Modelling Process, Environmental Impact Analysis: Philosophy and Methods[8] [*Overlay*].   This methodology employs an overlay technique via computer mapping. Data on a large number of environmental characteristics are collected and stored in the computer on a grid system of 1-km-square cells. Highway-route alternatives can be evaluated by the computer (by noting the impacts on intersected cells), or new alternatives may be generated via a program identifying the route of least impact.

The environmental characteristics used are rather comprehensive, particularly regarding land use and physiographic characteristics. Although the methodology was developed and applied to a highway setting, it is adaptable (with relatively small changes in characteristics) to other project types with geographically well-defined and concentrated impacts. Because the approach requires considerable amounts of data about the project region, it may be impractical for the analysis of programs of broad geographical scope. The manpower-skill, money, and computer-technology requirements of the approach may limit its application to major projects or to situations where a statewide computer data base exists (e.g., New York, Minnesota, Iowa).

Impact importance is estimated through the specification of subjective weights. Because the approach is computerized, the effects of several alternative weighting schemes can be readily analyzed.

The methodology is attractive from several viewpoints. It allows a demonstration of which weighted characteristics are central to a particular alternative route; it presents a readily understandable graphic representation of impacts and alternatives; it easily handles several subjective weighting systems; its incremental costs of considering or generating additional alternatives are low; and it fits well with developing regional and statewide data bank systems.

The mechanics of the approach (how impacts are measured and combined) may not be readily apparent from the reference cited. Considerable training beyond the information available in this reference may be required prior to using the approach.

A Procedure for Evaluating Environmental Impact[9] [*Matrix*].   This is an open-cell matrix approach identifying 100 project activities and 88 environmental characteristics or conditions. For each action involved in a project, the analyst evaluates the impact on every environmental characteristic in terms of impact magnitude and significance. These evaluations are subjectively determined by the analyst. Ecological and physical-chemical impacts are treated comprehensively; social and indirect impacts are discussed in part, and economic and secondary impacts are not addressed.

Because the assessments are subjective, resource requirements of the approach are very flexible. The approach was not developed in reference to any specific type of project and may be broadly applied with some alterations.

Guidelines for use of the approach are minimal, and several important ambiguities are likely in the definition and separation of impacts. The reliance upon subjective judgment, again without guidelines, reduces the replicability of the approach.

The approach is chiefly valuable as a means of identifying project impacts and as a display format for communicating results of an analysis.

Transportation and Environment: Synthesis for Action: Impact of National Environmental Policy Act of 1969 on the Department of Transportation[20] *[Checklist].* This approach is basically an overview discussion of the kinds of impacts that may be expected to occur from highway projects, and the measurement techniques that may be available to handle some of them. A comprehensive list of impact types and the stages of project development at which each may occur is presented. As broad categories, the impact types identified are useful for other projects as well as highways.

The approach suggests the separate consideration of an impact's amount, effect (public response), and value. Some suggestions are offered for measuring the amount of impact within each of seven general categories: noise, air quality, water quality, soil erosion, ecological, economic, and sociopolitical impacts.

Five possible approaches to handling impact significance are presented. Three of these are "passive" (requiring no agency action), such as "reliance upon the emergence of controversy." The other two involve the use of crude subjective weighting scales. No specific suggestions are made for the aggregation of impacts either within or between categories.

In general, the reference cited is a useful discussion of some of the important issues of impact analysis, particularly as they apply to transportation projects; however, it does not present a complete analytical technique.

A Comprehensive Highway Route-Selection Method,[11] *or Design with Nature*[12] *[Overlay].* This approach employs transparencies of environmental characteristics overlaid on a regional base map. Eleven to sixteen environmental and land use characteristics are mapped. The maps represent three levels of the characteristics, based upon "compatibility with the highway." While these references do not indicate how this compatibility is to be determined, available documentation is cited.

This approach is basically an earlier, noncomputerized version of the ideas presented in Krauskopf.[8] Its basic value is a method for screening alternative project sites or routes. Within this particular use, it is applicable to a variety of project types.

Limitations of the approach include its inability to quantify and identify possible impacts and its implicit weighting of all characteristics mapped.

Resource requirements of this approach are somewhat less demanding in terms of data than those of the Krauskopf approach, because information is not directly quantified, but rather, categorized into three levels. However, high degrees of skill and training are required to prepare the map overlays.

The approach seems most useful as a "first-cut method" of identifying and sifting out alternative project sites prior to preparing a detailed impact analysis.

A Methodology for Evaluating Manufacturing Environmental Impact Statements for Delaware's Coastal Zone[13] [*Matrix*]. This approach was not designed for impact analysis, although its principles could be adapted for such use. Employing a network approach, it links a list of manufacturing related activities to potential environmental alterations, major environmental effects, and finally, to human uses affected. The primary strength of the set of linked matrices is their utility for displaying cause-condition-effect networks and tracing out secondary impact chains.

Such networks are useful primarily for identifying impacts. The issues of impact magnitude and significance are addressed only in terms of high, moderate, low, or negligible damage. As a result of these subjective evaluations, the approach would have low replicability as an assessment technique. For such a use, guidelines would likely be needed to define the evaluation categories.

The approach incorporates indicators especially tailored to manufacturing facilities in a coastal zone, although most indicators would also be pertinent to other types of projects. It would perhaps be valuable as a visual summary of an impact analysis for communication to the public.

Environmental Resources Management[5] [*Matrix*]. This methodology employs a matrix approach to assess in simple terms the major and minor, direct and indirect impacts of certain water-related construction activities. It is designed primarily to measure only the physical impacts of water-resource projects in a watershed and is based upon an identification of the specific, small-scale component activities that are included in a project of any size. Restricted to physical impacts for nine types of watershed areas (e.g., wetlands) and 14 types of activities (e.g., tree removal), the procedure indicates four possible levels of impact-receptor interactions (major direct through minor indirect).

Low to moderate resources, in terms of time, money, and personnel, are required for this methodology, due principally to its simple method of quantification (major versus minor impact). However, the procedure is severely limited in its ability to compare different projects or the magnitude of different impacts.

Since there is no spatial or temporal differentiation, the full range of impacts cannot be readily assessed. Impact uncertainty and high-damage/low-probability impacts are not considered. Since only two levels of impact magnitude are identified, and the importance of the impacts is not assessed, moderate replicability results. The lack of objective evaluation criteria may produce fairly ambiguous results. NEPA requirements for impact assessments are not directly met by this procedure.

This methodology may be less valuable for actual assessment of the quantitative impacts of a potential project than for the "capability rating system," which determines recommended development policies on the basis of existing land characteristics. Thus, guidelines for desirable and undesirable activities, with respect to the nine types of watershed areas, are used to map a region in terms of the optimum land-use plan. The actual mapping procedure is not described; therefore, that aspect of the impact assessment methodology cannot be evaluated here.

Quantifying the Environmental Impact of Transportation Systems[16] *[Checklist]*. This approach, as developed for highway-route selection, is a checklist system based upon the concepts of probability and supply-demand. The approach attempts to identify the alternative with least social cost to environmental resources and maximum social benefit to system resources. Environmental resources elements are listed as: agriculture, wildlife conservation, interference noise, physical features, and replacement. System resources elements are listed as: aesthetics, cost, mode interface, and travel desired. Categories are defined for each element and used to classify zones of the project area. Numerical probabilities of supply and demand are then assigned to each zone for each element. These are multiplied to produce a "probability of least social cost" (or maximum social benefit). These "least social cost" probabilities are then multiplied across the elements to produce a total for the route alternative under examination.

The approach is tailored and perhaps limited to project situations requiring comparison of siting alternatives. While the range of environmental factors examined is limited, it presumably could be expanded to more adequately cover ecological, pollution, and social considerations.

Since procedures for determining supply and demand probabilities are not described, it is difficult to anticipate the amounts of data, manpower, and money required to use the approach. The primary limitations of this methodology are the difficulties inherent in assigning probabilities, particularly demand probabilities, and the implicitly equal weightings assigned each element when multiplying to yield an aggregate score for an alternative.

A Framework for Identification and Control of Resource Degradation and Conflict in the Multiple Use of the Coastal Zone[17]; *and Procedures for Regional Clearinghouse Review of Environmental Impact Statements— Phase Two*[18] [*Network*].   These two publications present a network approach usable for environmental impact analysis.   The approach is not a full methodology but rather a guide to identifying impacts.   Several potential uses of the California coastal zone are examined through networks relating uses to causal factors (project activities), to first order condition changes, to second and third order condition changes, and finally, to effects.   Major strength of the approach is its ability to identify the pathways by which both primary and secondary environmental impacts are produced.

The second reference also indicates data types relevant to each identified resource degradation element, although no specific measurable indicators are suggested.   In this reference, some general criteria suggested for identifying projects of regional significance are based upon project size and types of impacts generated, particularly land-use impacts.

Because the preparation of the required detailed networks is a major undertaking, the approach is presently limited to some commercial, residential, and transportation uses of the California coastal zone for which networks have been prepared.   An agency wishing to use the approach in other circumstances might develop the appropriate reference networks for subsequent environmental impact assessments.

Environmental Impact Assessment: A Procedure[19] [*Checklist*].   This methodology is a checklist procedure for a general quantitative evaluation of environmental impacts from development activities.   The type and range of these activities is not specified but is believed to be comprehensive.   The 50 impact parameters are sufficient to include nearly all possible effects and thereby allow much flexibility.   Subparameters indicate specific impacts, but there is no indication of how the individual measures are aggregated into a single parameter value.   While spatial differences in impacts are not indicated, both initial and future impacts are included and explicitly compared.

The moderate to heavy resource requirement, especially in terms of an interdisciplinary personnel team, increases as more subparameters are included and requires additional expertise in specific areas.   However, the actual measurements are not based on specific criteria and are only partially quantitative, having seven possible values ranging from on extremely beneficial impact to an extremely detrimental one.   Therefore, there may be room for ambiguous and subjective results, with only moderate replicability.

The assumption that impact areas are implicitly of equal importance allows aggregation of the results and project comparisons, but at the expense of realism. A specific methodology is mentioned for choosing the optimum alternatives in terms of the proportional significance of an impact vis-à-vis other potential alternatives. There is no explicit mention of either public involvement in the process or environmental risks.

The impact assessment procedure is presented as only one step in a total evaluation scheme, which includes concepts of dynamic ecological stability and other ideas. An actual description of the entire process is not indicated, however.

Guidelines for Implementing Principles and Standards for Multiobjective Planning of Water Resources[14] [*Checklist*]. This approach is an attempt to coordinate features of the Water Resources Council's Proposed Principles and Standards for Planning Water and Related Land Resources with requirements of NEPA. It develops a checklist of environmental components and categories organized in the same manner as the WRC Guidelines. The categories of potential impacts deal comprehensively with biological, physical, cultural and historical resources, and pollution factors, but do not treat social or economic impacts. Impacts are measured in quantitative terms wherever possible, and also rated subjectively on a "quality" and "human influence" basis. In addition, uniqueness and irreversability considerations are included where appropriate. Several suggestions for summary tables and bar graphs are offered as communications aids.

The approach is general enough to be widely applicable to various types of projects, although its impact categories are perhaps better tailored to rural than urban environments. While no specific data or other resources are required to conduct an analysis, an interdisciplinary project team is specified to assign the subjective weightings. Since quality, human influence, uniqueness and irreversibilities are all subjectively rated by general considerations, results produced by the approach may be highly variable. Significant ambiguities include a generally inadequate explanation of how human influence impacts are to be rated and interpreted.

Key ideas incorporated in the approach include explicit identification of the "without project" environment as distinct from present conditions, and a uniqueness rating system for evaluating quality and human influence (worst known, average, best known). The methodology is unique among those examined, because it does not label impacts as environmental benefits or costs, but only as impacts to be valued by others. The approach also argues against the aggregation of impacts.

Matrix Analysis of Alternatives for Water Resource Development[10] [*Checklist*]. Despite the title, this methodology can be considered to be a checklist under the definitions used here. Although a display matrix is used to summarize and compare the impacts of project alternatives, impacts are not linked to specific project actions. The approach was developed to deal specifically with reservoir-construction projects but could be readily adapted to other project types.

Potential impacts are identified within three broad objectives: environmental quality, human life quality, and economics. For each impact type identified, a series of factors is described to show possible measurable indicators. Impact magnitude is not measured in physical units but by a relative impact system. This system assigns the future state of an environmental characteristic without the project a score of zero; it then assigns the project alternative possessing the greatest impact on that characteristic a score of +5 (for positive impact), or −5 (for negative impact). The raw scores thus obtained are multiplied by weights determined subjectively by the impact analysis team.

Like the Georgia approach, [15] this methodology tests for the significance of differences between alternatives by introducing stochastic error factors and conducting repeated runs. The statistical manipulations are different from those used in the Georgia approach, however, and considered by Corps writers to be more valid.

Resource requirements of this methodology are variable. Since specific level types of data are not required, data needs are quite flexible. The consideration of error, however, requires specific skills and computer facilities.

Major limitations of the approach, aside from the required computerization, are the lack of clear guidelines about exactly how to measure impacts and the lack of guidance about how the future "no project" state is to be defined in the analysis. Without careful description of the assumptions made, replicability of analyses using this approach may be low, since only relative measures are used. Since all measurements are relative, it may be difficult to deal with impacts that are not clearly definable as gains or losses.

The key ideas of wider interest incorporated in this approach include reliance upon relative, rather than absolute, impact measurement, statistical tests of significance with error introduction, and specific use of the "no project" condition as a baseline for impact evaluation.

A Manual for Conducting Environmental Impact Studies[22] [*Checklist*]. This methodology is a checklist, unique in its almost total reliance upon social impact categories and strong public participation. The approach was developed for evaluating highway alternatives and identifies different impact analysis procedures for the conceptual, corridor, and design states of highway planning. All impacts are measured either by their dollar value or by a weighted function of the number of persons affected. (The weights used are to be determined subjectively by the study team.) The basis for most measurements is a personal interview with a representative of each facility or service impacted.

Resource requirements for such a technique are highly sensitive to project scale. The extensive interviewing required may make the approach impractical for many medium-sized or large projects, because agencies preparing impact statements seldom have the necessary manpower or money to contract for such an extensive interviewing.

Analyses produced by the approach may have very low replicability. This results from the lack of specific data used and the criticality of the decision regarding boundaries of the analysis, since many impacts are measured in numbers of people affected. There is also no means of systematically accounting for the extent to which these people are affected.

The key ideas of broader interest put forth by the approach are: the use of only social impacts, without direct consideration of physical impacts (e.g., pollution, ecology changes); the heavy dependence upon public involvement and specific suggestions about how the public may be involved; and the recognition of the need for different analyses of different project development stages.

Environmental Guidelines[4] [*Ad Hoc*].   The Environmental Guidelines are intended primarily as a planning tool for siting power-generation and power-transmission facilities. However, they address many of the concerns of environmental impact analysis and have been used to prepare impact statements. Viewed as an impact assessment methodology, the approach is an ad hoc procedure, suggesting general areas and types of impacts but not listing specific parameters to examine.

The approach considers a range of pollution, ecological, economic (business economics) and social impacts; however, it does not address secondary impacts, such as induced growth or energy use patterns. The format of the approach is an outline of considerations important to the selection of sites for each of several types of facilities (e.g., thermal generating plants, transmission lines, hydroelectrical and pumped storage, and substations). An additional section offers suggestions for a public information program.

Since the approach does not suggest specific means of measuring or evaluating impacts, no particular types of data or resources are required. The application of this approach is limited to the siting of electric power facilities with little carry-over to other project types.

5.6 FUTURE DIRECTIONS

This chapter has provided guidance for choosing an environmental impact assessment methodology, a description of six general categories of methodologies, a criteria for reviewing a given methodology to determine its weaknesses and strengths, a description of selected methodologies, and a reference listing of other methodologies, with a notation of the general category in which these methodologies can be classified. As mentioned previously in this chapter, depending upon the specific needs of the user and the type of project being undertaken, one particular methodology may be more useful than another. While it is possible to select one of the methodologies mentioned here for use by an agency to solve its specific needs for environmental impact analysis, no one methodology can effectively and economically be utilized for major agency programs. An agency, using the information and systems developed under existing methodologies, should investigate the feasibility of developing procedures and systems to address its specific needs for environmental impact analysis. In the long run, this can provide substantial cost savings and allow the agency to prepare meaningful and comprehensive EIA/EIS.

REFERENCES

1. Adkins, William G. and Burke Dock, Jr., *Interim Report: Social, Economic, and Environmental Factors in Highway Decision Making*, research conducted for the Texas Highway Department in cooperation with the U.S. Department of Transportation, Federal Highway Administration, Texas Transportation Institute, Texas A & M University, October 1971.

2. Dee, Norbert *et al.*, *Environmental Evaluation System for Water Resources Planning*, report to the U.S. Bureau of Reclamation, Battelle Memorial Institute, January 1972.

3. Dee, Norbert *et al.*, *Planning Methodology for Water Quality Management: Environmental Evaluation System*, Battelle Memorial Institute, July 1973.

4. *Environmental Guidelines*, Western Systems Coordinating Council, Environmental Committee, 1971.

5. *Environmental Resources Management*, prepared for the Department of Housing and Urban Development, Central New York Regional Planning and Development Board, October 1972.

6. Jain, R. K. *et al.*, *Environmental Impact Assessment Study for Army Military Programs*, Interim Report D-13/AD 771062, U.S. Army Construction Engineering Research Laboratory [CERL], December 1973.

7. Jain R. K. *et al.*, *Handbook for Environmental Impact Analysis*, Technical Report E-59/ADA006241, CERL, September 1974.

8. Krauskopf, Thomas M. and Dennis C. Bunde, *Evaluation of Environmental Impact Through a Computer Modelling Process*, Environmental Impact Analy-

sis: Philosophy and Methods, Robert Ditton and Thomas Goodale, eds., pp. 107–125, University of Wisconsin Sea Grant Program, 1972.

9. Leopold, Luna B. *et al.*, *A Procedure for Evaluating Environmental Impact*, Geological Survey Circular 645, Government Printing Office, 1971.

10. *Matrix Analysis of Alternatives for Water Resource Development*, U.S. Army Corps of Engineers, Tulsa District, July 31, 1972.

11. McHarg, Ian, *A Comprehensive Highway Route-Selection Method*, Highway Research Record No. 246, pp. 1–15, 1968.

12. McHarg, Ian, *Design with Nature*, pp. 31–41, Natural History Press, Garden City, New York, 1969.

13. Moore, John L. *et al.*, *A Methodology for Evaluating Manufacturing Environmental Impact Statements for Delaware's Coastal Zone*, report to the State of Delaware, Battelle Memorial Institute, June 1973.

14. Multiagency Task Force, *Guidelines for Implementing Principles and Standards for Multiobjective Planning of Water Resources*, U.S. Bureau of Reclamation, December 1972 [draft].

15. *Optimum Pathway Matrix Analysis Approach to the Environmental Decision Making Process: Test Case: Relative Impact of Proposed Highway Alternatives*, University of Georgia, Institute of Ecology, 1971.

16. Smith, William L., *Quantifying The Environmental Impact of Transportation Systems*, Van Doren-Hazard-Stallings-Schnacke, undated.

17. Sorensen, Jens, *A Framework for Identification and Control of Resource Degradation and Conflict in the Multiple Use of the Coastal Zone*, University of California, Berkeley, Department of Landscape Agriculture, 1970.

18. Sorensen, Jens and James E. Pepper, *Procedures for Regional Clearinghouse Review of Environmental Impact Statements—Phase Two*, Report to the Association of Bay Area Governments, April 1973.

19. Stover, Lloyd, V., *Environmental Impact Assessment: A Procedure*, Sanders and Thomas, Miami, Florida, 1972.

20. *Transportation and Environment: Synthesis for Action: Impact of National Environmental Policy Act of 1969 on the Department of Transportation 3*, prepared for Office of the Secretary.

21. Urban, L. V. *et al.*, *Computer-Aided Environmental Impact Analysis for Construction Activities: User Manual*, Technical Report E-50, CERL, March 1975.

22. Walton, L. Ellis, Jr. and James E. Lewis, *A Manual for Conducting Environmental Impact Studies*, Virginia Highway Research Council, January 1971.

23. Warner, M. L. and E. H. Preston, *A Review of Environmental Protection Impact Assessment Methodologies*, U.S. Environmental Protection Agency, April 1974.

6

Generalized Approach for Impact Analysis

Most federal agencies are large organizations with diversified activities and programs. To assess the environmental impact of implementing agency programs, it is necessary that a systematic approach be developed to relate agency activities to potential environmental impacts. It is not sufficient for the responsible agency to issue directives requiring participants and lower echelons to prepare environmental impact statements or assessments to meet the provisions of NEPA. Administrative support, funding for executing the NEPA requirements, and a systematic procedure with substantive guidance for preparing EIS's is needed for each agency. Impact assessment methodologies discussed in the previous chapter can assist in developing an agency-specific system.

A generalized approach for impact analysis system development for an agency is shown in Fig. 6.1.

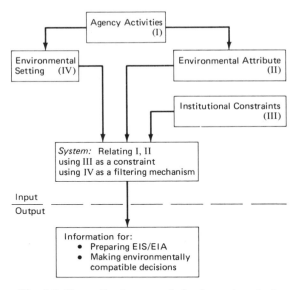

Fig. 6.1 Generalized approach for impact analysis.

6.1 AGENCY ACTIVITIES

In utilizing this generalized approach, the first thing one has to do is to categorize agency activities and actions such that these activities could be related to potential environmental impacts.   When categorizing agency activities, one has to intimately understand the various functions, programs, and operations of the agency and its components. The agency activities may be categorized into a hierarchical structure as shown below:

Functional area

Program

Subprogram

Basic activities

To provide the reader with an example of how this can be accomplished, the following paragraphs describe a case study for Army military programs.

Case Study.   In developing a methodology for relating Army activities to potential environmental impacts, it was necessary to develop a scheme for categorizing and classifying all Army activities in a systematic way. To develop a classification system, consideration was given to:[2]

1. Classification based on the Fiscal Code, as documented in Army regulations;

2. Classification of Army activities by installation; and
3. Classification based on the Army environmental impact guidelines.

It was recognized that, individually, each of the above approaches created unique problems regarding the scope and amount of detail required. For example, by inventorying only existing installations, the system would have been inflexible and would not have been capable of incorporating potential impacts in areas other than those specifically identified in the data base. New installations would then have to be totally assessed and entered as a specific addition to the data base. Also, in order to assess impacts at a specified installation, it would be necessary to review the baseline data for that specific site. Such information is usually not available in sufficient detail or appropriate format. Hence, specific installation review for development of basic activities associated with implementing Army programs was not possible.

Therefore, after active consultation with the potential users and careful review of agency guidance, a classification scheme was developed which synthesized the above approaches. This scheme generated the nine Army functional areas shown below:

1. Construction projects;
2. Operation, maintenance, and repair;
3. Training—basic to large-scale maneuvers;
4. Mission changes which increase or decrease the number or type of personnel at the installation or change the activities of people;
5. Real estate acquisition or outleases or disposal of land;
6. Procurement;
7. Industrial plants;
8. Research, development, test, and evaluation; and
9. Administration and support.

These functional areas are defined to encompass all Army activities. For each functional area, basic activities were identified. In most cases, the activities which were identified were at such a level of detail that it was necessary to relate them to the functional area through a hierarchy of activities. Therefore, for most functional areas, a hierarchy of Army activities was established as follows:

Functional area
Program
Subprogram
Aggregate activities
Detailed activities

Due to variations in the nature of the functional areas, some of the hierarchical levels were omitted in some functional areas. Further details concerning how the activities for Army military programs were developed are described in Jain *et al.*[2]

In addition to categorizing agency activities, it is necessary to develop a list of representative major actions and programs of an agency that might have a significant environmental impact, or whose impact, if implemented, might be considered controversial. As an example, a list for Army military programs is shown in Fig. C-2, Appendix C.

6.2   ENVIRONMENTAL ATTRIBUTES*

In order to relate agency activities to potential environmental impacts, it is desirable to categorize the elements of the environment into subsets. It should be recognized that the environment is a continuum and that there is interaction between the various environmental parameters. A minor impact on an environmental parameter could have more serious and far reaching secondary and indirect impacts on other parameters of the environment. For example, removal of vegetation may cause excessive soil erosion, which may cause excessive sediments in the receiving stream. This, in turn, will reduce the amount of sunlight that can penetrate the water, thus reducing the dissolved oxygen in the water. Dissolved oxygen plays an important role in the biological economy of water. Reduction of dissolved oxygen will adversely affect aquatic life and water quality of the stream.

The environmental attributes can be categorized into different subsets, depending upon the level of detailed information required and the particular needs of the agency. For example, it might be desirable to develop three different types and levels of attributes. These could be:

a. Detailed attribute
b. Review level attribute
c. Controversial attribute

The detailed attribute may be used to describe the conditions of the environment; any changes in the attribute would indicate changes in the environment. A review level attribute may provide an overview of the nature of the potential impacts. As such, this kind of parameter could be useful for summarizing the potential environmental impacts and provide synoptic information for personnel at the management level. Controversial attributes may be those parameters which, when impacted by the agency's activity, are likely to produce an adverse public reaction or controversy.

*May be referred to as environmental parameters, or environmental elements.

It is not sufficient to develop a list of environmental attributes. It is also necessary to give substance and meaning to these parameters by providing scientific information, such as definition of the attribute, how man's activity might affect this attribute, how this attribute can be measured and how this relates to other environmental attributes, etc. Information of this type for 49 selected attributes is included in Appendix B. These 49 attributes would correspond roughly to review level attributes—clearly the environment cannot be fully described using only this set of attributes.

Description

The following paragraphs give further information and delineate the items included in characterizing the attributes.

Definition of the Attribute. This item defines the environmental attribute. The definition also explains how the attribute relates to the environment.

Activities that Affect the Attribute. This item contains *examples* of human activities and suggests what type of activity affects the subject attribute.

Source of Effects. This item provides a brief discussion of some of the potential ways human activities will cause an impact on the subject environmental attribute.

It should be noted that these descriptions are intended to give the reader an overview of each attribute in the context of its role in impact analysis. None of the descriptions should be considered complete, as many of the individual subject areas themselves form the basis for complete texts. It is anticipated that familiarity with these 49 attributes can serve to expedite indisciplinary studies which frequently encounter difficulties due to lack of communication between disciplines. This communication problem can be overcome when the participants attain some understanding of each other's terminology, problems, and difficulties in achieving solutions to those problems.

Variables to be Measured. This item discusses the real world variables that are to be measured to indicate environmental impact. If necessary, the relationship of the measurement to the attribute is also discussed.

How Variables are Measured. This is one of the most important items in the attribute description. To the greatest extent possible, the methods for measuring impact on the variables are presented here. This includes information on sources of data that can be used to assist in measuring impact, primarily secondary data sources. References to additional technical materials that are required to adequately measure changes in the variables may be included. The types of skills that may be required in measuring impact on the variables are also discussed. For example, for collecting census data from published reports, no special skill is required; but for measuring sound levels, detailed technical capabilities may be required. The need for these capabilities is identified in this item. Special instruments for measuring impact, to the extent that they are required, are identified.

Evaluation and Interpretation of Data. When the data regarding impact have been collected, an additional step is required to determine whether the impact on the subject attribute is favorable or unfavorable. In addition, the evaluation of the severity of impact is also discussed. For some attributes, the method for converting the changes in the variable into another indicator of impact is presented. This permits comparison to other environmental attributes. Other attributes are not as easily evaluated, and evaluation of the impact may require considerable professional expertise.

Special Conditions. This item discusses the special measurement problems or difficulties that may be encountered in determining the impact on the subject attribute. These special conditions stem from poor availability of secondary data. If necessary, this item discusses the type and necessity for special measurement techniques. Examples of the special conditions would be the necessity for survey data regarding community values to provide baseline data for some of the impacts in the human environment category. Another example would be the need for extremely complicated measurement instruments, the use of which may require special expertise.

Geographical and Temporal Limitations. Discussed are the potential problems that might arise because of different geographical or time locations of impacts on the attribute. For example, many of the land attributes will have varying impacts, depending on the geographical location of the subject activity.

Mitigation of Impact. Each environmental attribute has the potential for being affected by human activities. However, it is also possible for the activities to be modified in such a way as to reduce the impact on the attributes. In this section, the methods for reducing impacts are discussed.

Secondary Effects. Impacts on other aspects or attributes of the environment may result in a secondary or an indirect manner. For example, an aircraft runway modification project may alter flight patterns, directly changing the sound levels in adjacent areas. These could lead to a shift in land use development, followed by a variety of biophysical and socioeconomic effects.

Other Comments. This item is reserved for information that does not fall within any of the other items relating to the environmental attributes.

Procedure for Using the Attribute Descriptor Package

The evaluation of environmental impact on an attribute-by-attribute basis, involves a straightforward review of each attribute description, keeping in mind the activity that may cause the impact. As the attribute is reviewed, the data collected, and the impacts identified, entries should be made in an environmental attribute list to indicate the potential impact of the human activity on the environment. A procedure for using the attribute descriptor package in the preparation of EIA/EIS is given in Appendix C.

6.3 INSTITUTIONAL CONSTRAINTS

Implementation of a project or action is subject to institutional constraints, such as emission standards for air quality control, effluent standards for wastewater discharge, and noise pressure levels for acceptable land uses. These institutional constraints could include federal, state, regional, or local environmental regulations, standards, or guidelines; as such, these could place severe constraints on the implementation of projects or actions. It is, therefore, important to carefully consider these institutional constraints in the environmental impact analysis process.

Since there are vast number of environmental regulations, and there is also the overlapping of agency jurisdictions, it is not always possible to obtain information regarding institutional constraints easily and expeditiously. To help solve this problem, many environmental legislative data systems have been developed. Described herein are some of the existing legislative data systems which the reader may want to use, depending upon his specific needs.

Legal Information Through Electronics (LITE)[1]

The LITE System is a "full-text" legislative retrieval system for the statutory decisional, treaty, and regulatory laws of the federal government. LITE performs searches for entire documents by recognizing user-selected keywords. This sytem has the following characteristics:

1. Only federal laws are included.
2. Full-text retrieval is comprehensive, but it is difficult to use easily for EIA purposes.
3. Legal language in the document may obscure the meaning and possible applicability of the legislation for nonlegal personnel.
4. The comprehensiveness of the search depends upon the choice of words for search items.

Justice Retrieval and Inquiry System (JURIS)

JURIS is a U.S. Department of Justice computer-based storage and retrieval system for federal statutory laws. It is specifically designed to serve federal lawyers and has some of the same characteristics as LITE: it covers only federal laws, and it is designed for legal personnel. Information from the system could be rather voluminous, and consequently, difficult to use for EIA purposes.

Computer-Aided Environmental Legislative Data System (CELDS)[5]

This system contains abstracts of environmental legislation and is designed for use in environmental impact analysis and environmental quality management. The abstracts are written in an informative narrative style, with all legal jargon and excessive verbiage removed. Characteristics of this system are:

1. Legislative information is indexed to a hierarchical keyword thesaurus, in addition to being indexed to a set of environmental attributes, which number approximately 700.
2. Information can be obtained for federal and individual state environmental legislation, as well as regulatory requirements related to the keywords or environmental attributes.
3. Appropriate reference documents, such as enactment/effective date, legislative reference, administrative agency, and bibliographical reference, are also included.
4. Information is structured in order to satisfy the user agency's (U.S. Army's) specific needs for environmental legislation; consequently, needs of other agencies may not be completely satisfied by this system.

Other Systems

Some proprietary systems for providing data on environmental legislation are in existence. These systems could be quite useful in providing the requisite information for an agency. Details on such systems can be obtained from the proprietors of the systems.

6.4   ENVIRONMENTAL SETTING

Depending upon the environmental setting (or environmental baseline) at a location where the project or action is to be implemented, severity or existence of the impact would vary. Consequently, when structuring a generalized EIA system, provisions need to be made for incorporating the site-specific environmental setting or baseline. In a systematic procedure, environmental baseline information serves as a quasifiltering mechanism, eliminating consideration of impacts unrelated to the specific site.

6.5 SYSTEM

After developing agency activities [I]*, appropriate environmental parameters [II], a system for obtaining relevant institutional constraints [III], and information on environmental baseline characteristics [IV], a system needs to be developed to relate I to II, using III as a constraint, and IV as a filtering mechanism. This system could be just as simple as a thinking process, a manual storage and manipulation system, or a computer-aided system. Rationale for utilizing a computer-aided system for such an analysis is discussed in Section 6-7.

6.6 OUTPUT

Output from such a system should be structured to provide information necessary for preparing an EIA/EIS and for making environmentally compatible management decisions. This output could include:

(a) An impact matrix, relative activities to potential environmental impacts;
(b) Abatement and mitigation techniques;
(c) Analytical cause-effect relationships providing quantitative information for some environmental areas;
(d) Institutional constraints which must be considered.

6.7 RATIONALE FOR A COMPUTER-BASED SYSTEM

As discussed previously, one of the options for systematizing the generalized approach for environmental impact analysis is to develop a computer-aided system. When one discusses utilization of computer-aided systems for environmental impact analysis, many questions arise, such as:

1. Can a meaningful computer-aided system be developed which is practical, useful, cost-effective, and still does not provide mechanical solutions to important environmental impact analysis problems?
2. Some environmentalists may question the development of any systematic procedure, computer-aided or otherwise, for environmental impact analysis.

*Refers to numbers in Fig. 6.1.

Before establishing a need for a computer-aided system, it might be well to look at some of the general problems associated with preparing an EIA/EIS. After discussions with agency personnel charged with preparing these documents, the following problems have been identified:

1. The excessive cost of preparing EIA/EIS;
2. The interdisciplinary expertise required by NEPA to prepare an EIA/EIS is not always available within the staffs of agencies;
3. Even with availability of interdisciplinary expertise, it is not always possible to determine secondary and cumulative impacts which would result from implementation of a given action. This means that additional fundamental research is needed to identify, in a meaningful way, the secondary and cumulative impacts of an action.
4. A vast amount of environmental information is scattered in various publications, reports, standards, and technical manuals. It is neither convenient nor economically feasible to scan all these information sources to make environmentally compatible decisions or to prepare an environmental impact asessment. It would not be economically feasible, for example, to obtain the necessary environmental regulatory information for preparing comprehensive EIA/EIS. For this reason alone, an efficient and cost-effective system for storing and accessing data is needed. This requirement leads, almost inevitably, to a computer-aided system.
5. For some environmental impact analysis problems, it is necessary to develop cause-effect analytical models. It would not be possible to operate these analytical models economically without the aid of computer systems.

To address the above cited problems, a computer-aided system may be the answer. A computer-aided system does not imply a mechanical system which would solve complex environmental problems mechanistically, but rather a system that would provide a tool to allow the user to address these problems in a comprehensive and systematic manner. One such system, called the Environmental Impact Computer System (EICS)[2,3,4] is being developed at the U.S. Army Construction Engineering Research Laboratory (CERL).

REFERENCES

1. *JAG Law Review* **XIV**, *No. 1:* 10–13, 25–34, 35–67 (1972).

2. Jain, R. K. *et al.*, *Environmental Impact Assessment Study for Army Military Programs*, Interim Report D-13/AD771062 (U.S. Army Construction Engineering Research Laboratory [CERL], December 1973).

3. Lee, E. Y. S. *et al.*, *Environmental Impact Computer System*, Technical Report E-37 (CERL, September 1974).

4. Urban, L. V. *et al.*, *User Manual—Computer-Aided Environmental Impact Analysis for Construction Activities*, Technical Report E-50 (CERL, March 1975).

5. Webster, R. D. *et al.*, *Development of The Environmental Technical Information*, Interim Report E-52 (CERL, April 1975).

7

Procedures for Reviewing Environmental Impact Statements

Any formal Federal agency document prepared at one level usually requires review at higher levels. This is also the case for environmental impact statements (EIS). In fact, EIS's are reviewed at many different levels within the proponent agency, and also, these documents are reviewed by other Federal and state agencies with jurisdiction by law or special expertise with respect to environmental impacts. Reviews of these documents are also made by conservation and environmental groups and concerned members of the community, especially those who might be affected by the implementation of the project or the action.

In view of the involvement of persons at various levels and organizations in the review of EIS documents, presented here is a framework for a procedure for reviewing these documents for construction-related projects. The framework and examples for construction-related projects are given for illustrative purposes. Review procedures for actions and projects other than construction can also be developed using the discussion in this chapter as a guide.

A review procedure can be used both by the reviewer and the preparer of an EIS document for ascertaining completeness, accuracy, and validity of the document. However, it should be kept in mind that as new requirements for the EIS documents are levied and as environmental concerns include new areas, such as energy and resource conservation, the review procedure would also require updating to meet the new demands.

In general, a review procedure should allow the reviewer to (a) ascertain the completeness of the EIS document; (b) assess the validity and accuracy of the information presented; and (c) become familiar with the project very quickly and ask substantive questions to determine whether any part of the document needs additional work and/or strengthening.

7.1 AN EXAMPLE REVIEW PROCEDURE

The characteristics of the project must be known and must be combined with a set of screening questions, shown in Table 7.1. These questions broadly categorize construction projects by their characteristics according to extent of potential impacts. The response rating of these questions are recorded along with the response score. Example response ratings are shown in Table 7.2 and may be used to guide the determination of the appropriate response rating and associated score. The scores may then be summed for the project to provide a total score. The score provides a rationale to categorize construction project impacts into three major levels (I, II and III). Next, the detailed EIS review criteria for the three major project levels are used to review the document. Levels I, II and III are presented in Tables 7.3, 7.4, and 7.5, respectively.

Project Screening Questions

Twelve project screening questions in Table 7.1 have been developed to categorize potential project impacts according to project characteristics. The questions cover a broad range of major environmental impacts associated with the construction projects. These questions are answered either by "Yes" or "No," or by "High," "Medium," or "Low." Determination of an answer is based upon response rating criteria.

Response Rating Criteria

Specific criteria were developed to determine the answer to each project screeening question. Such criteria prescribe what is meant by a "High," "Medium," or "Low" (or "Yes" or "No") rating for a particular question.

TABLE 7.1 Screening Questions

No.	Question	Rating	Score
1.	What is the approximate cost of the construction project?	High Medium Low	10 5 0
2.	How large is the area affected by the construction project?	High Medium Low	10 5 0
3.	Will there be a large, industrial type of project under construction?	Yes No	10 0
4.	Will there be a large, water-related construction activity?	Yes No	10 0
5.	Will there be a significant waste discharge (in terms of quantity and quality) to natural waters?	Yes No	10 0
6.	Will there be a significant disposal of solid waste (quantity and composition) on land as a result of construction and operation of the project?	Yes No	10 0
7.	Will there be significant emissions (quantity and quality) to the air as a result of construction and operation of the project?	Yes No	10 0
8.	How large is the affected population?	High Medium Low	10 5 0
9.	Will the project affect any unique resources (geological/historical/archaeological/cultural/ecological)?	Yes No	10 0
10.	Will the construction be on floodplains?	Yes No	10 0
11.	Will the construction and operation be incompatible with adjoining land use in terms of aesthetics/noise/odor/general acceptance?	Yes No	10 0
12.	Can the existing community infrastructure handle the new demands placed upon it during construction and operation of the project (roads/utilities/health services/vocational education/other services)?	No Yes	10 0

Example rating criteria presented in Table 7.2 for each screening question were developed by use of informed professional judgment and were meant to apply to construction projects. Suggested response rating criteria shown in Table 7.2 would have to be modified to apply to other types of projects and as experience in their use shows shortcomings.

Project Screening Criteria

Each response rating from Table 7.2 is assigned a score of 10, 5, or 0. For each "Yes," a project gets a score of 10; for each "No," the score is 0; for "High," "Medium," or "Low" ratings, scores assigned are 10, 5, and 0, respectively.

Possible total scores for all combinations of various construction projects range from 0 to 120. Within this range, the following three levels of projects are defined:

Level I:	Small-impact projects	Scores	$0 - \leqslant 60$
Level II:	Medium-impact projects	Scores	$> 60 - \leqslant 100$
Level III:	High-impact projects	Scores	> 100

These levels (or other appropriate ranges) may be used to discriminate between projects that require detailed versus less detailed review. The potentially high-impact project should be given the most thorough review, while the others should be given a less intensive review.

Review Criteria

Review criteria are employed to assess the completeness and accuracy of the impact statement. The review level is established by the score of the project screening exercise.

Review criteria for levels I, II and III are presented in Tables 7.3, 7.4, and 7.5, respectively. The criteria are divided into completeness and accuracy categories. *Completeness* criteria are meant to assess whether or not potential environmental impacts considered in the statement meet NEPA's full disclosure requirements necessary for project evaluation. *Accuracy* criteria are designed to assess the correctness and reliability of relevant information included in the document.

Figure 7.1 illustrates this review procedure in a flow chart.

7.2   ILLUSTRATION OF REVIEW PROCEDURE

The following example illustrates use of the review procedure.

TABLE 7.2   Example Response Rating Criteria

No.	Criteria	Rating
1.(a)	The construction is less than or equal to $1 million.	Low
1.(b)	The construction cost is >$1 million but <$20 million.	Medium
1.(c)	The construction cost is >$20 million.	High
2.(a)	The area affected by construction is less than or equal to 10 acres.	Low
2.(b)	The area affected by construction is >10 acres and ≤50 acres.	Medium
2.(c)	The area affected by construction is >50 acres.	High
3.(a)	An industrial-type project costing more than $1 million is involved.	Yes
3.(b)	Otherwise.*	No
4.(a)	The large water-related construction project consists of one or more of the following: A dam A dredging operation of 5 miles or longer; disposal of dredged spoils A bank encroachment that reduces the channel width by 5 percent Filling of a marsh slough ≥5 acres Continuous filling of 20 or more acres of riverine or estuarine marshes A bridge across a major river (span: 400 feet).	Yes
4.(b)	Otherwise.	No
5.(a)(1)	At least one of the following waste materials is discharged into the natural streams: Asbestos PCB Heavy metals Pesticides Cyanides Radioactive substances Other hazardous materials (specify)	Yes
5.(a)(2)	Rock slides and soil erosion into streams may occur because: No underpinning is specified for unstable landforms No sluice boxes/retention boxes/retention basins are specified for excavation and filling.	Yes
5.(b)	Otherwise.	No
6.(a)(1)	At least one of the following solid wastes is disposed on land: Asbestos PCB Heavy metals Pesticides Cyanides Radioactive substances Other hazardous materials.	Yes

No.	Criteria	Rating
6.(a)(2)	The solid waste generated is greater than 2 pounds/capita/day.	Yes
6.(b)	Otherwise.	No
7.(a)(1)	If there are to be: Concrete aggregate plants—EIS does not specify dust control devices.	Yes
7.(a)(2)	Hauling operations—EIS does no specify use of dust control measures.	Yes
7.(a)(3)	Road grading or land clearing—EIS does not specify water or chemical dust control.	Yes
7.(a)(4)	Open burning—EIS does not specify disposal of debris.	Yes
7.(a)(5)	Unpaved roads—EIS does not specify paved roads on construction sites.	Yes
7.(a)(6)	Asphalt plants—EIS does not specify proper dust control devices.	Yes
7.(b)	Otherwise.	No
8.(a)	Less than 20 persons are displaced by the project.	Low
8.(b)	From 20 to 50 persons are displaced by the project.	Medium
8.(c)	More than 50 persons are displaced by the project.	High
9.(a)(1)	A rich mineral deposit is located on the construction site.	Yes
9.(a)(2)	A historical site or building is located at or near the construction site.	Yes
9.(a)(3)	An existing or potential archaeological site is located near the construction project.	Yes
9.(a)(4)	A rare or endangered species is resident on the land proposed for construction.	Yes
9.(b)	Otherwise.	No
10.(a)	The construction project is on a 100-year floodplain.	Yes
10.(b)	Otherwise.	No
11.(a)(1)	No visual screening is specified in the EIS for the construction site.	Yes
11.(a)(2)	No progressive reclamation of quarry and/or disposal sites is proposed.	Yes
11.(a)(3)	No permissible noise level specifications are stated for vibrators, pumps, compressors, pile-drivers, saws, and paving breakers.	Yes
11.(b)	Otherwise.	No
12.(a)	The projected demand for community services exceeds existing or planned capacity. These services include: Water supply Wastewater treatment/disposal Electric generation	Yes

TABLE 7.2   (Continued)

No.	Criteria	Rating

	Transportation
	   Educational and vocational facilities
	   Cultural/recreational facilities
	   Health care facilities
	   Welfare services
	   Safety services; fire, flood, etc.	

12.(b)	Otherwise.	No

*"Otherwise" implies that none of the previously mentioned situations are applicable to the project.

TABLE 7.3   Level I—Review Criteria

I. COMPLETENESS
  A. Project Description
     1. Purpose of action   1. Identify and list major purposes of the projects.
                                  2. Describe each purpose in narrative form.
     2. Description of action   1. Identify and list names of major project activities
                                  2. Provide a brief description of these activities.
                                  3. Show location of activities on an appropriate scale map.
     3. Environmental setting   1. Describe the environmental quality prior to proposed action based upon macro factors such as:
                                      Air
                                      Water
                                      Land
                                  2. Describe existing federal activities (projects and operations) and their impacts.
  B. Land Use Relationships
     1. Conformity and conflicts   1. Identify existing federal, state, and local plans, policies, and controls.
                                    2. Discuss the extent of the project's conformity and conflicts with:
                                    Air quality plan under the Clean Air Act
                                    Water quality plan under the Federal Water Pollution Control Act Amendments of 1972
                                    Other plans
     2. Conflict assessment   1. Discuss the extent of conflict reconciliation.
                                    2. Discuss reasons for incomplete reconciliation.
  C. Impacts
     1. Positive and negative effects 1. Discuss the project's national impact upon:
                                    Air
                                  Water
                                  Land
                                  2. Discuss the project's international impact upon the above factors.

I. COMPLETENESS
  C. Impacts
    2. Direct/indirect consequences

1. Discuss the project's primary effects upon the environment.
2. Discuss the project's secondary effects upon the environment.

  D. Alternatives

1. Identify and describe alternatives.
2. Discuss benefits and costs.

  E. Unavoidable Effects

1. Discuss unavoidable impacts.
2. State techniques chosen to mitigate adverse impacts.

  F. Short-Term Uses Versus Long-Term Productivity

1. Identify short-term gains and losses from the project.
2. Identify long-term gains and losses from the project.
3. Discuss possible trade-offs between short-term uses versus long-term productivity.
4. Identify foreclosure of future options.

  G. Commitment of Resources

1. Identify major resources needed to construct the project.
2. Identify major resources that will be irreversibly used (i.e., they will cease to exist when the project is completed):
    Labor
    Materials
    Natural features
    Cultural objects

  H. Other Interests

1. Identify selected benefits of the proposed action, based upon terms of the present and projected needs of the proponent agency.
2. Identify selected benefits of alternatives, based upon the above mentioned criteria.
3. Discuss briefly the mitigation effects of selected benefits.

  I. Discussion of Review

1. Summarize comments of other agencies.
2. Identify all points of disagreement.
3. Indicate agency's response to the issues raised and comments made by the reviewers.

II. ACCURACY
  A. Readability

1. Write clearly.
2. Remove all ambiguities.
3. Avoid use of technical jargon; all technical terms should be clearly explained.

TABLE 7.3   (Continued)

II. ACCURACY
   B. Flavor and Focus
   1. Do not slant or misinterpret findings.
   2. Avoid use of value-imparting adjectives or phrases.
   3. Avoid confusion or mix-up between economic, environmental, and ecological impacts and productivity.
   4. Avoid unsubstantiated generalities.

   C. Quantification
   1. Use well-defined, acceptable qualitative terms.
   2. Quantity factors, effects, uses, and activities that are readily amenable to quantification.

   D. Data
   1. Identify sources of information.
   2. Use data consistent with official data; if data are unofficial, explain why.

   E. Methods and Procedures
   1. It is not necessary to use detailed models and quantitative techniques.
   2. Identify and describe methods and techniques used.

   F. Professional Expertise
   1. Use data collection and analysis procedures designed by recognized professional experts.
   2. Identify scientists participating in the EIS preparation.
   3. State qualifications of these scientists.
   4. Use expert judgment for all high-level impact areas where there is uncertainty.

   G. Interpretation of Findings
   1. Consider all high impact features before they are eliminated.
   2. Identify and discuss all controversial points.
   3. Discuss knowledge limitations.
   4. Identify uncertain findings.
   5. Give reasons for not selecting alternatives.

TABLE 7.4   Level II—Review Criteria

I. COMPLETENESS
   A. Project Description
      1. Purpose of action
      1. Identify all purposes of the project.
      2. Briefly describe each purpose.
      3. Explain the extent to which each purpose is served, quantifying on the basis of available data.

I. COMPLETENESS
  A. Project Description

2. Description of action	1. Identify and list all project activities.
	2. Describe all activities clearly and in detail.
	3. Relate each activity to the project's purpose.
	4. Provide a location map on an appropriate scale.
	5. Describe structures to be built.
	6. Give area covered by each structure.
	7. Discuss important design features.
	8. Explain construction practices.
3. Environmental setting	1. Describe environmental quality prior to proposed action based upon:

        Air
          Physical/chemical quality
          Health effects
        Water
          Physical/chemical quality
          Ecology
          Aesthetic quality
        Land
          Land-use pattern
          Soil erosion
          Aesthetic quality
          Noise conditions
          Ecology

2. Describe existing federal activities (projects and operations) and their impacts upon:
        Air
        Water
        Land
3. Describe existing state and local activities and their impacts upon:
        Air
        Water
        Land

  B. Land-Use Relationships

1. Conformity and conflicts	1. Identify existing federal, state, regional, and local plans, policies, and controls.
	2. Identify potential plans, policies, and controls.
	3. Identify various conflicts and conformities of project characteristics, based upon existing and potential criteria defined above.
	4. Assess, on the basis of the above criteria, the extent of conformity and conflict of the project with:

        Air quality plan
        Water quality plan
        Noise levels and impact on land uses
        Transportation plan

TABLE 7.4 *(Continued)*

I. COMPLETENESS
  B. Land-Use Relationships

1. Conformity and conflicts	Regional land use plan
	Community services plan
	Other plans
	5. Briefly describe the conflicts and conformities.
2. Conflict assessment	1. Define alternatives for conflict reconciliation.
	2. Establish the extent of conflict reconciliation for each alternative.
	3. Justify the choice of reconciliation alternatives.
	4. Discuss reasons for incomplete reconciliation.

  C. Impacts

1. Positive and negative effects	1. Quantify, if possible, national and regional impacts of the project upon:
	Air
	Physical/chemical quality
	Health effects
	Water
	Physical/chemical quality
	Ecology
	Aesthetic quality
	Land
	Land-use pattern
	Soil erosion
	Aesthetic quality
	Noise conditions
	Ecology
	2. Quantify, if possible, international impacts upon the above factors.
	3. Assess, synthesize, and document the impacts.
2. Direct/indirect consequences	1. Quantify, if possible, primary effects upon:
	Human health
	Ecology
	Vegetation
	Materials
	2. Quantify, if possible, secondary effects upon:
	Labor productivity
	Food chain
	Unemployment
	Alienation, etc.
	3. Describe the significance of the above effects.

  D. Alternatives

	1. Identify a set of selected alternatives in terms of:
	Location
	Equipment
	Operation procedure
	Engineering design
	2. Assess costs of each alternative.
	3. Assess benefits of each alternative.
	4. Identify feasible alternatives.

I. COMPLETENESS

   E. Unavoidable Effects

       1. Identify unavoidable impacts.
       2. Quantify above impacts wherever possible.
       3. Develop a profile of such impacts.
       4. Identify mitigation alternatives.
       5. Assess costs and benefits of these alternatives.
       6. Select a set of feasible alternatives.

   F. Short-Term Uses Versus Long-Term Productivity

       1. Identify short-term gains and losses from the project.
       2. Quantify the above gains and losses where possible.
       3. Identify long-term gains and losses from the project.
       4. Quantify the above gains and losses where possible.
       5. Assess trade-offs betwen short-term losses and long-term losses.
       6. Assess trade-offs between short-term losses and long-term gains.
       7. Develop a profile for trade-offs.
       8. Identify future options without the proposed project.
       9. Identify options that will be foreclosed by the project.

   G. Commitment of Resources

       1. Identify all resources needed to construct the project.
       2. Quantify major resources that will be used.
       3. Quantify major resources that will cease to exist when the project is completed:
          Labor
          Material
          Natural features
          Cultural objects
       4. Classify the above resources into major categories.

   H. Other Interests

       1. Quantify major benefits of the proposed action, based upon terms of the present and projected needs of the proponent agency.
       2. Quantify major benefits of alternatives based upon the above mentioned criteria.
       3. Discuss in detail the mitigation effects of major benefits.
       4. Establish superiority of proposed action.

   I. Discussion of Review

       1. Summarize comments of other agencies.
       2. Identify all points of disagreement.
       3. Indicate agency's response to the issues raised and comments made by the reviewers.

TABLE 7.4   (Continued)

II. ACCURACY
  A. Readability
        1. Write clearly.
        2. Remove all ambiguities.
        3. Avoid use of technical jargon; all technical terms should be clearly explained.

  B. Flavor and Focus
        1. Do not slant or misinterpret findings.
        2. Avoid use of value-imparting adjectives or phrases.
        3. Avoid confusion or mix-up between economic, environmental, and ecological impacts and productivity.
        4. Avoid unsubstantiated generalities.

  C. Quantification
        1. Use well-defined, acceptable qualitative terms.
        2. Quantify factors, effects, uses, and activities that are readily amenable to quantification.

  D. Data
        1. Identify all sources.
        2. Describe procedures for data collection.
        3. Be consistent with official data.
        4. Give reasons for use of unofficial data.

  E. Methods and Procedures
        1. Use quantitative techniques where data and knowledge allow.
        2. Identify and describe all techniques used.
        3. Identify estimates, based upon expert judgment.
        4. Use methods that are acceptable to professionals.

  F. Professional Expertise
        1. Use data collection and analysis procedures designed by recognized professional experts.
        2. Identify scientists participating in the EIS preparation.
        3. State qualifications of these scientists.
        4. Use expert judgment for all high-level impact areas where there is uncertainty.

  G. Interpretation of Findings
        1. Consider and discuss all impact areas before any are dismissed as not applicable.
        2. Discuss areas where there is uncertainty.
        3. Consider the range of possibilities for uncertain areas.
        4. Analyze alternatives in detail, and give reasons for not selecting them.
        5. Identify and describe all controversial points.

TABLE 7.5   Level III—Review Criteria

I. COMPLETENESS
   A. Project Description
      1. Purpose of action
         1. Identify all purposes of the project.
         2. Describe each purpose in detail.
         3. Establish criteria for measuring achievement of each purpose.
         4. Quantify the extent to which each purpose is achieved.
         5. Use innovative quantification approaches.
         6. Develop a purpose achievement profile.
      2. Description of action
         1. Identify and list all project activities.
         2. Describe all activities clearly and in detail.
         3. Relate each activity to the project's purposes.
         4. Provide a location map on an appropriate scale.
         5. Provide maps and sketches of all layouts and complete engineering drawings.
         6. Provide detailed engineering plans.
         7. Describe structures to be built.
         8. Provide construction details about:
               Timing
               Material flow
               Equipment use
               Construction monitoring plans
               Safety plans
               Labor needed
               Services required
               Pollution abatement procedures and hardware
         9. Give the area covered by each structure.
        10. Discuss important design features.
        11. Explain construction practices.
        12. Develop contingency plans for rare events which might occur during implementation of the project.
        13. Assess in detail the methods for shipping of materials and equipment.
      3. Environmental setting
         1. Describe the environmental quality prior to proposed action based upon:
               Air
                  Particulates
                  Sulfur dioxide
                  Hydrocarbons
                  Carbon monoxide
                  Nitrogen oxide
                  Oxidants
               Water
                  Physical/chemical quality
                     Flow pattern
                     Turbidity

TABLE 7.5 *(Continued)*

I. COMPLETENESS
  A. Project Description
    3. Environmental setting

Dissolved oxygen
pH
Phosphates
Hazardous substances
Ecology
  Habitat
  Diversity
  Endangered species
Aesthetic quality
  Appearance
  Land/water interface
  Floating materials
  Wooded shoreline
Land
  Specific parameters of land-use patterns
  Soil erosion, aesthetic quality
  Noise conditions
  Ecology

2. Describe existing federal activities and their impact upon the above parameters.
3. Discuss existing state and local activities and their impacts upon the above parameters.

  B. Land-Use Relationships
    1. Conformity and conflicts

1. Identify existing federal, state, regional, and local plans, policies, and controls.
2. Identify potential plans, policies, and controls.
3. Define quantitative requirements for conformity and conflict criteria.
4. Identify various conflicts and conformities, using a matrix approach or project characteristics, based upon the existing and potential criteria defined above.
5. Quantify, on the basis of the above criteria, the extent of the project's conformity and conflict with:
  Air quality plan
  Water quality plan
  Noise levels and impact on land uses
  Transportation plan
  Regional land-use plan
  Community services plan
  Other plans
6. Discuss conflicts and conformities in detail.
7. Specifically state the conflicts that need to be reconciled.

    2. Conflict assessment

1. Define alternatives for conflict reconciliation.
2. Quantify the extent of conflict reconciliation for each alternative for the proposed project.

I. COMPLETENESS
  B. Land-Use Relationships
    2. Conflict assessment

3. Systematically compare the overall conflict reconciliation for the various alternatives.
4. Select and justify the choice of alternative reconciliation.
5. Quantify the extent of incomplete reconciliation.
6. Justify incomplete conflict reconciliation.

  C. Impacts
    1. Positive and negative effects

1. Quantify, as well as possible, national and regional impacts of the project upon:
  Air
    Particulates
    Sulfur dioxide
    Hydrocarbons
    Carbon monoxide
    Nitrogen oxide
    Oxidants
  Water
    Physical/chemical quality
     Flow pattern
     Turbidity
     Dissolved oxygen
     pH
     Phosphates
     Hazardous substances
    Ecology
     Habitat
     Diversity
     Endangered species
    Aesthetic quality
     Appearance
     Land/water interface
     Floating materials
     Wooded shoreline
  Land
    Specific parameters of land-use patterns
    Soil erosion
    Aesthetic quality
    Noise
    Ecology
4. Quantify as much as possible the potential international impacts upon the above factors.
5. Assess, evaluate, and document the impacts listed above.

    2. Direct/indirect consequences

1. Quantify as much as possible, in terms of specific, measurable parameters, all primary effects upon:
  Human health
  Ecology
  Vegetation
  Materials

TABLE 7.5 (Continued)

I. COMPLETENESS
  C. Impacts
    2. Direct/indirect consequences

    2. Quantify, as much as possible, in terms of specific, measurable parameters, all secondary effects upon:
      Labor productivity
      Food chain
      Unemployment
      Alienation, etc.
    3. Develop specific profiles for each impact.

  D. Alternatives

    1. Identify a set of possible alternatives in terms of:
      Location
      Equipment
      Operation procedure
      Engineering design
    2. Quantify costs of each alternative.
    3. Quantify benefits of each alternative.
    4. Compare benefits with costs.
    5. Establish the most desirable alternatives.
    6. Develop a profile for the desirable alternatives, their benefits, and their costs.

  E. Unavoidable Effects

    1. Identify unavoidable impacts.
    2. Quantify the above impacts as well as possible.
    3. Develop a profile of unavoidable impacts.
    4. Justify why the unavoidable impacts cannot be eliminated.
    5. Identify mitigation alternatives.
    6. Quantify, as well as possible, the costs and benefits of these alternatives.
    7. Compare the benefits and costs of these alternatives.
    8. Select a set of feasible alternatives.
    9. Develop a profile for the unavoidable effects of these alternatives.

  F. Short-Term Uses Versus Long-Term Productivity

    1. Identify short-term gains and losses from the project.
    2. Quantify the above gains and losses as well as possible.
    3. Identify long-term gains and losses from the project.
    4. Quantify the above gains and losses as well as possible.
    5. Assess trade-offs between short-term gains and long-term losses.

I. COMPLETENESS
 F. Short-Term Uses Verses Long-
    Term Productivity

    6. Assess trade-offs between short-term losses
       and long-term gains.
    7. Develop a profile for trade-offs.
    8. Identify and describe future options without
       the proposed project.
    9. Identify and describe options that will be
       foreclosed by the project.

 G. Committment of Resources

    1. Identify all resources needed to construct the
       project.
    2. Quantify all resources that will be used.
    3. Quantify all resources that will cease to exist
       when the project is completed:
       Labor
       Material
       Natural features
       Cultural objects
    4. Classify the above resources into major
       categories.
    5. Develop a profile for irreversible commitment
       of resources.

 H. Other Interests

    1. Quantify all benefits of the proposed action
       based upon terms of the present and projected
       needs of the proponent agency.
    2. Quantify all benefits of alternatives based
       upon the above mentioned criteria.
    3. Discuss in detail the mitigation effect of major
       benefits.
    4. Establish superiority of proposed action.

 I. Discussion of Review

    1. Summarize comments of other agencies.
    2. Identify all points of disagreement.
    3. Indicate agency's response to the issues raised
       and comments made by the reviewers.

II. ACCURACY
 A. Readability

    1. Write clearly.
    2. Remove all ambiguities.
    3. Avoid use of technical jargon; all technical
       terms should be clearly explained.

 B. Flavor and Focus

    1. Do not slant or misinterpret findings.
    2. Avoid use of value-imparting adjectives or
       phrases.
    3. Avoid confusion or mix-up between eco-
       nomic, environmental, and ecological impacts
       and productivity.
    4. Avoid unsubstantiated generalities.

TABLE 7.5 (*Continued*)

II. ACCURACY
    C. Quantification

    1. Use well-defined, acceptable qualitative terms.
    2. Quantify factors, effects, uses, and activities that are readily amenable to qualification.

    D. Data

    1. Identify all sources.
    2. Use up-to-date data.
    3. Use field data collection programs as necessary.
    4. Use technically-approved data collection procedures.
    5. Give reasons for use of unofficial data.

    E. Methods and Procedures

    1. Use quantitative estimation procedures, techniques, and models for arrival at the best estimates.
    2. Identify and describe all procedures and models used.
    3. Identify sources of all judgments.
    4. Use procedures and models accpetable by professional standards.

    F. Professional Expertise

    1. Use data collection and analysis procedures designed by recognized professional experts.
    2. Identify scientists participating in the EIS preparation.
    3. State qualifications of these scientists.
    4. Use expert judgment for all high-level impact areas where there is uncertainty.

    G. Interpretation of Findings

    1. Consider and discuss all impact areas before any are dismissed as not applicable.
    2. Give thorough treatment to all controversial issues, and discuss the implications of all results.
    3. Consider the implications for each area of a range of outcomes having significant uncertainty.
    4. Analyze each alternative in detail and give reasons for not selecting it.
    5. Scrutinize and justify all interpretations, procedures, and findings that must stand up under expert professional scrutiny.

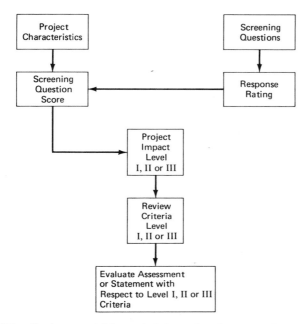

Fig. 7.1.   Environmental impact statement review procedure process.

Project Description

A federally funded family housing project of 300 units is to be built on 30 acres of land at a cost of $6.7 million. Occupants will move to the housing project from the community. The project site is located on gently sloping terrain, bounded by two intermittently flowing streams. Confluence of these streams occurs near one end of the site, forming a second stream, which flows directly into a small pond approximately 2600 ft from the site. Timber wastes from the construction site will be disposed by burning at the site. An asphalt batching plant will be operated on-site. Dust control measures will be employed on haul roads. Adjacent land use consists of family housing. Additional educational and family services will be required in the new location.

Screening Questions and Response Rating

Tables 7.1 and 7.2 are used to determine the level at which the review should be conducted.

QUESTION 1:   The cost of the project is $6.7 million. This corresponds to a rating of "Medium."

No.	Question	Rating	Score
1.	What is the approximate cost of the construction project?	High / (Medium) / Low	10 / (5) ← / 0
2.	How large is the area affected by the construction project?	High / (Medium) / Low	10 / (5) ← / 0
3.	Will there be a large, industrial type of project under construction?	Yes / (No)	10 / (0)
4.	Will there be a large, water-related construction work?	Yes / (No)	10 / (0)
5.	Will there be a significant waste discharge (in terms of quantity and quality) to natural waters?	Yes / (No)	10 / (0)
6.	Will there be a significant disposal of solid waste (quantity and composition) on land as a result of construction and operation of the project?	Yes / (No)	10 / (0)
7.	Will there be significant emissions (quantity and quality) to the air as a result of construction and operation of the project?	(Yes) / No	(10) ← / 0
8.	How large is the affected population?	(High) / Medium / Low	(10) ← / 5 / 0
9.	Will the project affect any unique resources (geological/historical/archaeological/cultural/ecological)?	Yes / (No)	10 / (0)
10.	Will the construction be on floodplains?	Yes / (No)	10 / (0)
11.	Will the construction and operation be incompatible with adjoining land use in terms of aesthetics/noise/odor/general acceptance?	(Yes) / No	(10) ← / 0
12.	Can the existing community infrastructure handle the new demands placed upon it during construction and operation of the project (roads/utilities/health services/vocational education/other services)?	(No) / Yes	(10) ← / 0
		TOTAL	50

Fig. 7.2   Example scoring of screening questions.

QUESTION 2:   The area affected is 30 acres, and this corresponds to a rating of "Medium."

QUESTION 3:   There is no industrial involvement, so the rating is "No."

QUESTION 4:   There is no large, water-related construction, according to criteria listed in Part 4(a), Table 7.2; the rating is "No."

QUESTION 5:   None of the waste materials listed in Part 5(a)(1), Table 2, are discharged, nor will rock slides or soil erosions occur; the rating is "No."

QUESTION 6:   None of the criteria in Parts 6(a)(1) and 6(a)(2), Table 2 apply; the rating is "No."

QUESTION 7:   Although dust control on haul roads is to be used, asphalt plant control devices are not specified. In addition, there will be open burning. The rating is "Yes."

QUESTION 8:   The construction of 300 units will generate a displacement of far more than 50 persons from the community to a new location. The rating is "High."

QUESTION 9:   Although not specified in the project description given above, none of the criteria in Parts 9(a)(1-4) apply in this case, so the rating is "No."

QUESTION 10:   Construction is not on a floodplain, so the rating is "No."

QUESTION 11:   Occupants of family housing adjacent to the project site are likely to be inconvenienced by noise, dust, and aesthetic problems. The rating is "Yes."

QUESTION 12:   The addition of 300 families to the new site will increase the need for educational, family, and other services in the area. The rating is "No."

Figure 7.2 is a copy of Table 7.1 in which the ratings and scores have been circled according to the proper response. Adding the scores gives a value of 50; therefore, the document review should be conducted at Level I—Small Impact Projects. The review criteria of Level I in Table 7.3 should be used.

7.3 SUMMARY

In order to assist administrative and staff personnel, this chapter has presented procedures for reviewing EIS for construction-related projects. These review procedures allow the preparer and user to:

(a) assess the completeness of the EIS, and
(b) ascertain the validity and accuracy of the information included in the EIS document.

In addition, by using these procedures, the reviewer can become familiar with the project very quickly and ask substantive questions to determine whether any part of the EIA/EIS document needs additional work or strengthening.

This chapter has described the essential elements of a review procedure.

This procedure provides a comprehensive and systematic approach for reviewing EIS documents for construction-related projects. An example illustrating use of this procedure to review an EIS for a family housing construction project was included to further clarify the procedure. It is expected that this methodology will be expanded and refined, based upon experience gained from application to actual EIS documents.

The review procedure presented in this chapter was first suggested in 1975[1] and was designed to be used in conjunction with the output provided by the Environmental Impact Computer System (EICS).[2,3]

REFERENCES

1. Jain, R. K., N. L. Drobny, and S. Chatterjee, *Procedures for Reviewing Environmental Impact Assessments and Statements for Construction Projects*, Technical Report, Construction Engineering Research Laboratory [CERL], August 1975.

2. Jain R. K., T. A. Lewis, L. V. Urban, and H. E. Balbach, *Environmental Impact Assessment Study for Military Programs*, Technical Report D-13, CERL, December 1973.

3. Urban, L. V., H. E. Balbach, R. K. Jain, E. W. Novak, and R. E. Riggins, *Computer-Aided Environmental Impact Analysis for Construction Activities: User Manual*, Technical Report E-50, CERL, April 1975.

8

Special Issues

The list of potential special issues that might be dealt with in environmental impact analysis is very large. This chapter focuses on three of the more central special issues that individuals, groups, or agencies conducting impact analyses must address. The need for greater public participation to facilitate the acceptability of impact analyses has resulted in a variety of actions, some of which are more harmful than useful in promoting participation and understanding by the public. The historical development and rationale for public participation is discussed and a preliminary framework for analysis of techniques is presented. The theoretical underpinnings of public participation methods range from political theory (democracy and the representative state) to information theory (learning curves and information costs). In addition, the value of a decision being made with full public knowledge and participation will be substantially different from one made without public involvement.

In another issue area, the question of economics and its appropriate relationship to environmental impact analysis has been raised for almost all impact assessments and statements. Some argue that economics does not belong in environmental impact analysis at all and others feel that it is an integral part of the analysis. The debates that have occurred will not be reviewed in this text. Instead, the section on economic impact analysis is presented to elaborate on a rationale for including it in an environmental impact analysis. In addition, several approaches for performing the analysis are discussed.

The next subject in this section, energy and the environment, rounds out what some people refer to as the three E's, environment, economics, and energy. Our energy resources clearly represent an irreplaceable environmental asset. The production of energy from raw materials poses serious environmental questions: emissions into the atmosphere, damming of rivers, long-term storage of nuclear wastes, and the disruption of land forms in stripping coal reserves; to name a few. Compliance with environmental regulations may further increase the energy per capita consumption of the United States, already far ahead of other industrialized nations. The section on energy discusses some of the difficult issues and trade-offs that must be considered as our concern for the environment is manifested in reduced environmental insult.

8.1 PUBLIC PARTICIPATION

Virtually all large public sector projects have significant effects on the community within which they are to take place. Generally, public agencies are charged with the responsibility of acting on behalf of the constituency they serve or represent. Actions that require environmental impact assessments and statements are usually extensive and are likely to affect the local community in a variety of ways, and these effects may be good or bad. However, the need for the project to take place in response to the requirements of the local community establishes the necessity for effective public participation. Without such participation, the project may take on a direction or emphasis that (although seemingly directed toward public benefit) is counterproductive to the communities' needs.

We are all familiar with the epithet that "taxation without representation" is unfair, unjust, uncalled for, not desirable, and not in the best interests of the subject population. Similarly, public sector activities in our interests that evolve without our inputs to guide direction, quality, and quantity also seem equally misguided.

Early American Experiences in Public Participation

One of the early contemporary studies of public involvement in decision-making is Thomas Jefferson Westerbaker's *The Puritan Oligarchy—The Founding of American Civilization* (1947). Westerbaker points out the obvious conflict between the Jeffersonian concept of a participatory democracy with the reality of the church society in Massachusetts. From its inception, the Massachusetts Bible State exemplified the government of the many by the few, represented in the comparatively small body of church members.

The theoretical political base of the United States and most other democratic governments accepts, as one of its central tenets, the Jeffersonian concept of participatory democracy. This concept establishes the need for political figures to seek the consent of the governed when making decisions affecting the welfare of the state and its citizens.

This theory finds classic expression in the town meeting and assumes the educability of the citizen public, the predominance of reason, the availability of full information, and free access to the decision making process, with the end product being understanding, consensus, harmony, and sound decisions.

James Madison, however, recognized the basic incongruity of this concept and wrote:

> Those who hold and those who are without property have ever formed distinct interest in society. Those who are creditors, and those who are debtors, fall under a like discrimination. A landed interest, a manufacturing interest, a mercantile interest, a moneyed interest, with many lesser interests, grow up of necessity in civilized nations, and divide them into different classes, actuated by different sentiments and views. The regulation of these various and interfering interests forms the principal task of modern legislation and involves the spirit of party and faction in the necessary and ordinary operations of government.

> *Tenth Federalist Paper*

Problems associated with this "principal task of modern legislation" to respond equally to various "publics" have been re-articulated many times since Madison's attempt. Much of this problem revolves around the question of citizen involvement in governmental decision making and has resulted in great difficulty identifying and defining pragmatic approaches to operationalize American government.

Eighteenth and nineteenth centuries were dominated by the frontier. Settlers, in those centuries, perceived the American continent as both a savage wilderness which should be conquered, and as the new world, full of inexhaustible resources of every kind. So, basically, the destiny of man appeared at the time to be to tame the wilderness and exploit its resources.

In the nineteenth century, conservationists were philosophers and not activists. For example, Henry David Thoreau quietly and eloquently recorded in his journal his conviction that preservation is a worthwhile goal and that wilderness is justified by the inspiration that men can draw from it.[24] Men like Thoreau were out of the mainstream of the commercial and political life of the nation. They had little impact on its policies. For them, preservation of natural amenities was an ethical and a moral issue. It appears that their philosophical ideas had little practical influence on the real problem, but what their writings did provide were philosophical foundations for the next generation of conservationists.

These philosophical concepts proved to be insufficient to persuade the public. For example, the city of San Francisco proposed to create a reservoir in the spectacular Hetch-Hatchy Valley in Yosemite National Park. The question was whether a man-made empoundment should be built within a national park. Other sites were available, but the Hetch-Hatchy site was the least costly.[24]

John Muir, founder of the Sierra Club, proponent of wilderness, argued that the reservoir would be inconsistent with the national park concept. Also, he argued that it would consume a magnificent scenic area and would offer no recreational benefits. Muir's philosophical and ethical arguments proved to be insufficient when put against the economics-based arguments of the proponents. In 1913, the Hetch-Hatchy reservoir was approved by the Congress.[24]

In the early 1950's, environmentalists and conservationists, in addition to arguing for preservation as a philosophical concept, utilized engineering and hydrologic studies to support their views. The case in point was the Echo Park Dam in Western Colorado. As a result of the arguments set forth by the conservationists, and as a result of public participation and involvement, this particular project was dropped from the development plans.[24]

More recently, in the late 60's and early 70's, environmentalists have extensively used mass media, public education campaigns, and also the provisions of NEPA. Consequently, many Federal actions have had to be either modified or dropped for environmental reasons.

Citizens' role in environmental preservation started as an ethical and philosophical concern of few; then it was supported by economic and engineering scientific arguments; and more recently, mass media, public education campaigns, and the provisions of NEPA and CEQ guidelines have been used by citizens in their crusade to preserve and protect the environment.

Contemporary Experience in Public Participation

Layer upon layer of administrative and management agencies isolate citizens from the decisions that affect their environment. This isolation places citizens

in a restrictive position. Either they must approach environmental management agencies to request assistance in dealing with a problem, or they may
demand solutions to a problem through the judicial process. In both cases,
the citizen is responding to administrative decisions.

Rather than only respond to decisions, there is a need to involve more
citizens in the decision making process itself. This approach increases citizens'
presence in the administrative agencies. It also reduces the need for antagonistic and legalistic behavior by the citizens.[1] Citizen protest and legal battles
are expensive and time-consuming alternatives to involving citizen groups
in the planning process.

The rational for encouraging more and better public participation is rooted
in a variety of theories that cut across a range of social sciences. But it should
be recognized at the outset that just as there may be a rationale for improving
public participation in decision making, so, also, there is a time, place, and
reason for restricting public inputs into the decision making process. This
arises primarily because the time involved in making a decision that conforms with public desires may involve costs that exceed the benefits of greater
public acceptability.

Public participation is required by NEPA, Presidents Executive Order
11514, and the latest CEQ guidelines. CEQ encourages federal agencies to
use innovative techniques beyond the public hearing for effective public
participation early in the planning stages of federal projects and actions.
CEQ suggests many administrative techniques that federal agencies can use
to obtain such participation; for example, an early warning or early notice
system for informing the public of upcoming federal actions in which they
may want to provide input. In addition, comments received by the federal
agency on the draft EIS are required to be carefully evaluated and considered
in the decision making process.

Effective Community Participation

Effective community participation has been defined as a community acting
with full information, equal access to decision-making institutions, and implementing its jointly articulated objectives.[2] Based on this definition, several
important objectives should be achieved to attain effective public
participation.

First, there must be as much information as possible made available to
the public. There often is considerable misinformation about the nature
of most proposed projects, even when they do not involve withholding of
information. This lack of communication precludes effective citizen participation in many cases.

Agencies often allow their image as a public spirited service institution
to be maligned because organizations and individuals construe the agencies'

failure to provide adequate information as cavalier or inconsiderate. If, instead, an active program of public information and public participation were undertaken, not only would there be more useful public input, and therefore a better project, but there would probably be less criticism of the agency.

Second, the community members must have access to the decision process. Allowing or encouraging community involvement in problem identification and discussion, without influence on the ultimate decision, is not an answer to the problem—rather, it becomes a charade.

Third, for community participation to be effective, the input provided by citizenry should result in a course of action consistent with their desires and with the needs of their fellow community members. The agency must have the power to act on behalf of the citizens, and the decision must reflect the joint objectives of the agency and the community.

The elements of an effective participation system are:

(a)   information exchange
(b)   access to decision-making
(c)   implementation powers.

Various types of communication exchanging provide for the elements of an effective program. For a communication technique to function as a public participation tool, it must allow for citizens to become involved in decision making. This definition means that techniques that only allow one-way communication, such as newspaper articles, are not very useful. Newspaper articles may, however, be one prerequisite communication step in a public participation program that includes other forms of interaction with a well-informed public. A wide range of techniques contain some or all of the characteristics necessary for a public participation program.

Figure 8.1 presents a list of selected techniques for public participation and communication. This list may be used as an aid in determining which techniques are best suited to particular planning programs.[2] It must be recognized that a comprehensive and operational community involvement program would be composed of a variety of these communication techniques.

Benefits from an Effective Public Participation Program

The catalog of reasons why decision making should not be made in a public forum but should reside in a central locus is extensive. Centralized decision making leads to more rapid, cost-effective, decisive decisions, permitting effective and efficient leadership. The military is built on this decision making mode. Congress seems to act on issues lethargically, appearing to be inefficient and ineffective in comparison with the Executive branch. However, this slow action has benefits; it provides an opportunity for diverse views to be accommodated. This perspective on the value of public participation suggests that decisions made on behalf of the public by centralized agencies can be substantially enhanced by providing channels for public input.

There is a greater likelihood that more viable or innovative alternatives to a project will be identified by opening up the process to the public. Community members are well aware of their own resources, limitations (most often), and problems. The diverse perspectives of the community's citizens provide input that could otherwise be obtained only through extensive field work by the agency sponsoring the project. There is, further, the possibility that there might be a closer integration of planning and development with existing area planning efforts in which major input has already been made by the public. A community may be expected to react unfavorably when previous input to other pertinent plans is summarily disregarded by agency planners and decision makers.

Active public involvement may also insure that the final product, which the community helps to develop, will be successfully implemented. Implementation is much more likely where the community has taken an active concern in planning problems, and has played an important role in generating and evaluating alternative solutions. An important spinoff from a positive program of public involvement is a positive public attitude toward not only the proposed project, but toward the agency as well.

8.2 ECONOMIC IMPACT ANALYSIS

Economic impact analysis is one component of environmental impact analysis that is frequently misunderstood. The relevance of "economics" as an element of the "environment" is difficult to rationalize, particularly when economics has been set forth as an equal and opposite factor to be traded off against the environment. However, just as the ecological structure within which a project is to take place determines the effect that project will have on the environment, so the economic structure within which a project is to take place will affect the environment. This generalization requires that the reader perceive the environment in its broadest sense, to cover all of the factors that affect the quality of a person's life. This quality is determined by all the factors contributing to health and welfare, both for short and long term periods. A general list of factors that describe the environment in this context includes both ambient physical conditions such as air, water, land, noise, etc., and the existing social, political, and economic structure of a community. The economic structure *per se* is an environmental factor that might be affected just as is air or water.

Measurement of effects on the economic structure have traditionally been mixed with the effects on the economic conditions of a community where a project is taking place. As a result, consequences of projects such as changes in employment, income, and wealth for a community have been used to

	Communication Characteristics						Objectives of Education and Participation Techniques					
Techniques for Communicating and Involving the Public	Degree of Public Contact Achieved	Degree of Impact on Decision Makers	Degree of User Sophistication	Ease of Use and Preparation	Ability to Respond to Varied Interests	Degree of Two-Way Communication	Inform/Educate	Identify Problems and Values	Get Ideas/Solve Problems	Feedback	Resolve Conflict/Research Consensus	Implement Solutions
Group A—Large Group Meetings												
1—Public Hearings	2*	1	2	2	0	0	1	2		1		
2—Public Meetings	2	1	2	2	0	1	1	2		1		
Group B—Small Group Meetings												
3—Presentations to Community Groups	1	3	2	2	3	3	2	2		2		
4—Field Trips and Site Visits	1	2	3	2	2	3	2	3		2		
5—Advisory Body	1	3	1	2	3	3	3	3	2	3	2	
6—Task Force	1	1	1	2	3	3	3	3	3	3	2	
7—Gaming and Role Playing Exercises	1	1	2	3	2	3	2	3	2		2	2
8—Values Clarification Exercises	1	1	2	3	2	3	2	3	2		2	3
9—Workshops and Seminars	1	1	1	2	3	3	2	3	3	2	2	
10—Delphi Exercises	2	3	2	1	3	3	1	2	3	3	3	

	C1	C2	C3	C4	C5	C6	C7	C8	C9	C10
Group C—Organizational Approaches										
11—Regional and/or Local Offices	2	2	3	3	2	1	2	3	2	3
12—Citizen Representation on Policy Boards	1	3	1	2	1	2	3	3	2	2
13—Ombudsman and Community Advocate	1	2	3	3	3	2	3	3	2	2
14—Public Interest Center	3	1	3	3	2	2	2	2	2	2
Group D—Media										
15—Information Pamphlets, Brochures, and Summary Reports	2	1	2	2	1	1	1	2	1	
16—Slide and Film Presentations	2	1	3	3	1	1		3		
17—Tape Recorded Information Network	2	1	3	2	2	2	2	2	2	
18—Radio Talk Shows	3	1	3	1	3	2		1		
19—Press Releases and News Letters	3	1	2	2	1	1		1		
Group E—One-to-One Communications										
20—Response to Public Inquiries	1	1	1	2	2	1				
21—Attitude Surveys—Mailed, Telephone, and Personal Interviews	2	2	3	3	3	2	3	3	2	3
Group F—Legal Mechanisms										
22—Citizen Suits	1	3	3	2	3	3	3	3	3	3
23—Environmental Impact Review Statement	1	1	1	3	1	2	3	3	3	2

*These evaluations are based on 1, 2, and 3, respectively, representing poor, average, and good.

Fig. 8.1   Capabilities of environmental resources education and public participation techniques.*

describe the economic aspects of environmental impact. Economic conditions are generally not directly part of an environmental impact assessment. However, economic structure and environmental consequences of changes in conditions are part of an environmental impact assessment. Economic conditions *are* the factors, however, that should be balanced against environmental gains and losses. Thus, economic factors may be divided into two categories; the first relating to a description of the economic structure, and the second a description of economic conditions.

> Structure:
>   employment by industry
>   public versus private sector income
>   economic base
>   income/wealth distribution
> Conditions:
>   income
>   employment
>   wealth
>   production/output

Economics plays another important role in defining an environmental impact because changes in economic conditions lead to effects on the environment. Increases (or decreases) in income, production, or output lead to changes in effluents from production and consumption of goods and services. Changes in effluents of various types affect the environment. Therefore, there are secondary effects on other environmental attributes from changes in economic conditions.

Direct observation of economic structure and conditions is difficult. Because of this, it is usually necessary to use a model of the economic system to estimate and project variables. Models that are used are constructed so that changes resulting from project activity can be traced through to the effect on the economic variables of structure and conditions.

Project activity is the force (exogenous) that drives the economic model, as shown in Fig. 8.2. The model estimates results in impacts on economic

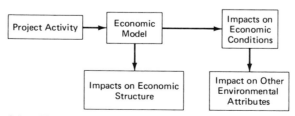

Fig. 8.2.   The relationship between project activity and impacts.

conditions and/or structure. The changes in economic conditions are translated, usually through another model, into impacts on other environmental attributes.

Economic Models

In the schema in Fig. 8.2, the economic model plays an important role in estimating and projecting the effects of a project. There are several types of models that might be employed in this framework to help in estimating the effects of project activities on the environment. Two of these models, the input-output model and the economic base model, are discussed next.

Input-Output Model

The study of economics and its relationship to environmental quality has most frequently been approached by analyzing environmental considerations separately from economic considerations. Individual environmental factors such as air, water, and solid waste have also been treated separately from one another. As Ayres and Kneese noted, "the partial equilibrium approach is both theoretically and empirically convenient, but ignores the possibility of important trade-offs between the various forms in which materials may be discharged back to the environment."[3] Recent attempts at model development have recognized the limited value of this partial perspective. Isard analyzes the economic and ecologic linkages based on a linear flow model. The Isard model requires a detailed matrix of ecologic resource flows to describe all of the interrelated processes that take place within the ecosystem.[4] Cumberland developed a model that adds rows and columns to the traditional input-output table to identify environmental benefits and costs associated with economic activity, and to distribute these costs by sectors.[5] Leontief's general equilibrium model is an extension of his fundamental economic input-output formulation, in which the model assumes one additional sector in the basic input-output table.[6] Pollution generated by the economy is consumed, at a cost, by an antipollution industry, represented by this additional sector.

An important modification of Leontief's approach was developed by Laurent and Hite.[7] This model is composed of an inter-industry matrix, a local use matrix, an export matrix, and an ecological matrix. For each economic sector, it shows the physical environmental change in terms of natural resources consumed and pollutant emission rates per dollar of output. These effects are computed by deriving the Leontief inverse of the inter-industry matrix and multiplying the environmental matrices by that inverse.

This section discusses an extension of this approach to environmental impact analysis.

A regional anaysis model based on a standard input-output table may be expanded to incorporate industrial land use and natural resource requirements as well as pollutant waste characteristics of industry into the I/O table.[8] The regional model may be viewed as a standard inter-industry input-output matrix that has been supplemented with land use, natural resource, and emission sectors where

R—Resource matrix specifying land and other resource requirements of each sector.

P—Pollution matrix specifying the non-marketed by-products of each sector.

A—Input-output table including resource and pollution sectors.

The regional analysis model may be expressed as follows:

$$A = RP (I - A)^{-1}.$$

In applying this model, the Leontief inverse $(I - A)^{-1}$ is calculated and the land-use, natural-resource, and pollutant matrix is multiplied by the inverse. This calculation provides an estimate of the impact of a proposed project

TABLE 8.1   Land-Use/Natural-Resource Inputs and Pollution Emissions by Sector

		SIC$_1$ . . . . SIC$_z$
Natural-Resource Inputs	Total land area (ft^2/employee)	
	Floor space (ft^2/employee)	
	Parking area (ft^2/employee)	
	Building site area (ft^2/employee)	
	Domestic water (gals/$ output)	
	Cooling water (gals/$ output)	
	Process water (gals/$ output)	
	Total water (gals/$ output)	
	$\vdots$	
	n	
Pollutant emissions	Particulates (lbs/$ output)	
	Sulfur dioxide (lbs/$ output)	
	Water discharge (gals/$ output)	
	5-day BOD (lb/$ output)	
	Solid waste (cu yd/$ output)	
	$\vdots$	
	m	

on the land use changes, natural resource, and waste generation characteristics of the region.

Reasonably good data are available for applying this model. An extensive survey of industrial land use requirements was completed by the Bureau of Public Roads, U.S. Department of Transportation, in July 1970, including (1) total land area, (2) floor space, (3) parking area, and (4) building site area by Standard Industrial Classification (SIC) category.[9] Land area per employee and floor area per employee are viewed in this model as resource requirements.[10]

The data on land use/natural resource inputs and waste emission characteristics may be organized in matrix form as shown in Table 8.1. Specific information must be collected to derive environmental coefficients for water

TABLE 8.2   Total Economic and Environmental Impacts Generated by Adding 600 New Employees—An Example

	Cotton Finishing Plant (Sector 2261)	Fabricated Structural Steel Plant (Sector 3441)
Economic Factors		
Value added by industry	7,982,000	8,761,000
Employment opportunity	2,046	2,118
Land Use and Natural Resources		
Domestic water (gal)	291	317
Cooling water (gal)	4,771	8,235
Process water (gal)	15,023	11,979
Total water intake (gal)	16,938	17,665
Land area (sq ft)	14,300,350	14,728,435
Floor space (sq ft)	1,073,721	1,173,006
Parking area (sq ft)	1,291,594	1,622,903
Building site (sq ft)	754,078	879,064
Waste Emissions		
Particulates (lb)	2,710,845	4,166,001
Hydrocarbons (lb)	1,205,817	1,328,205
Sulfur dioxide (lb)	147,225	164,735
Gaseous fluoride (lb)	0	0
Hydrogen sulfide (lb)	15,997	16,976
CO_2 (lb)	87,382	104,641
Aldehydes (lb)	3,481	3,861
NO_2 (lb)	54,887	61,561
Discharge (gal)	12,031	9,453
5 day BOD (lb)	1,395,944	1,023,066
Suspended solids (lb)	930,809	592,683
Solid waste (cu yd)	53,231	56,835

and land input requirements; and air, water, and other pollutant output emissions.

Applying the model to analyze the impact of specific project activity (adding new employees to specific economic sectors) produces output illustrated by the example in Table 8.2. The major advantage of this model is that it produces the detailed information necessary to analyze the effect of project activity on the environment, both in terms of the structure of the economy and in terms of the secondary effects of changed economic activity on the environment. The main disadvantage is that it is relatively expensive to operate because, for reliability, some primary data collection is essential.

Economic Base Model

Another approach to modeling the economic elements of environmental impact analysis is the Economic Impact Forecast System (EIFS), developed for use in assessing the effects of military projects.[14] This model is based on the principle that the total effect of an injection of new money in an economy can be estimated by determining how much of the money remains in the economy and is re-spent, and how much is removed from circulation.

The principal objective of EIFS is to answer the question, "What would happen to the local economy if certain activities affecting the economy were to take place?" To answer this question, the participants of the local economy must be described and their inter-reactions presented.

The three basic participants in the local economy are local government, households, and business. Local households purchase some goods and services from local business, receive wages and profits from the sale of their productive services, and pay taxes and consume services provided by local government. Local businesses sell goods and services, purchase inputs, pay taxes, and also receive services from local government. Local government purchases goods and services from business, purchase inputs such as labor from households, collects taxes, and provides public goods and services such as police protection, fire protection, and libraries. Thus, it can be seen that there is significant interrelationship between the various elements of the economy.

The effect of one household in the economy obtaining additional money can be traced using this model.

In the model, the flow of this money, as it works its way through the various sectors, is traced. Part of the money received would be put into the household's savings, and the rest would be used to finance purchases. Some of the products that are purchased would be purchased locally; others would be purchased from other regions. Purchases that are made from other regions require dollars to flow out of the local economy, while money received by local business would be used locally to hire labor, purchase products, pay taxes, and become profits. The wages received by local labor would, in turn, be partly saved and partly spent. Some of the products purchased from labor income would have been produced in the subject region; others would have been produced elsewhere. Thus, the cycle repeats until the original injection is completely dissipated.

The general idea is that money injected into a local economy would be partly retained and re-spent in the area and partly dissipated into other regions. The total effect of the initial injection depends upon many factors, but the sum total will be greater than the initial injection, that is, the initial injection will have a multiple effect upon the local economy. This concept is called the multiplier effect and is extremely important to the assessment of impacts.

Any change in injections into the economy will consequently lead to a multiple change in income. The EIFS system assumes that, in the short run, the variable most likely to change is exports. As a result, exports are considered *basic* to economic growth. Other activities in a region are nonbasic, in the sense that they do not result in any money inflows, at least not under the assumptions made about the short-run model.

If the relationships postulated in the multiplier analysis are constant, the multiplier can be written as:

$$\frac{1}{1-S} = \frac{1}{1 - \dfrac{\text{nonbasic income}}{\text{total income}}} = \frac{1}{\dfrac{\text{basic income}}{\text{total income}}} = \frac{\text{total income}}{\text{basic income}}$$

where

S = the proportion of total income attributable to nonbasic economic activity.

An estimate of the proportion of total income of the region, based upon export sales or basic industry sales, is necessary to use this multiplier. Fortunately, there are many techniques that can be used for an indirect estimation at low cost. The central assumption of indirect techniques is that there is a fixed relationship between the export industries in a region and the other local businesses. Perhaps the most widely used method to isolate export industries is the *location quotient* technique.[15]

Location quotients are based upon a comparison of regional employment with national employment. Because the United States is basically self-sufficient, if a region has a greater percent of its employment in a particular industry than does the nation, it is assumed to be specialized in the production of that commodity. Producing an excess of its own requirements, the region must export that commodity to other regions. A hypothetical example of the calculation of location quotients is given in Table 8.3.

Next to each industry grouping is the percent of the total national employment that an industry contains. In the next column is given the total employment in the region for each industry, and the percent of total regional employment that industry contains is calculated in column 3. The location quotient is derived by dividing column 3 by column 1. A location quotient greater than 1 indicates that the subject region exports that commodity to other regions. Location quotients less than 1 imply that the good is not produced locally in sufficient quantities to satisfy local needs, and hence, must be imported.

Given that basic industries have been identified, how is employment in that industry allocated to exports? In column 5, the location quotient minus one is divided by the original location quotient. This provides an estimate of the percent of employment in the industry that is involved in export activity. Multiplying column 5 by column 2 provides the estimate of the number of export employees for each industry. The multiplier is simply the ratio of export employment to total regional employment. In this example, the multiplier would be 5, indicating that a $1 increase in export demand would cause regional income to change by $5.

The multiplier concept is the basis for the development of this model. The details of the model take this general concept and use it to convert project activity (usually in dollars) into changes in business and economic activity. The strength of this approach is that results can be obtained relatively quickly and inexpensively. The major weakness is that the results are presented primarily in terms of changes in economic conditions, and changes in terms of structure or secondary effects on the environment from the changed economic activity are not dealt with in this approach.

Future Direction for Economic Impact Analysis

Both of the models discussed in this section are operational and have been applied in specific impact analyses. Thus, they represent applied approaches to dealing with the economic aspects of environmental impact analysis. The first approach can be used to develop detailed estimates of changes in structure and the secondary impacts in the local economy, while the second approach provides a broad estimate of the effect on economic conditions in a community where changes have been introduced by project activity.

The areas that require more work are in the development of models that predict changes in economic structure and better methods for esti-

TABLE 8.3   Location Quotients for a Hypothetical Region

Industry or Sector	Percent of National Employment	Regional Employment	Percent of Regional Employment	Location Quotient	$\dfrac{LQ-1}{LQ}$	No. of Export Employee
Services	.40	400	.40	1.00	—	
Durable goods manufactured	.20	75	.075	.375	—	
Nondurable manufactured	.10	25	.025	.25	—	
Trade	.30	500	.50	1.667	.40	200
Total		1000				

$$\text{Multiplier} = \frac{\text{total employment}}{\text{basic employment}} = \frac{1{,}000}{200} = 5$$

mating the pollutants that are generated by economic activity. In addition, the extent to which changes in structure are "good" or "bad" needs to be assessed.

8.3 ENERGY AND IMPACT ANALYSIS

Critics of the environmental protection movement have been quick to point accusing fingers at "environmentalists," blaming them (at least in part) for the energy crisis. While many of these claims are unfounded, it should be recognized that many interrelationships indeed exist—the production and use of energy results in environmental consequences, and environmental protection measures also have effects on energy use patterns. For example, the shift away from coal following the 1970 Clean Air Act Amendments and the delays in nuclear power plant construction have resulted in increases in oil and gas demands. Other examples frequently cited include decreased gasoline mileage due to emission control requirements for new automobiles and delays in Alaskan pipeline construction and offshore drilling efforts.

U.S. energy problems reached crisis proportion with the Arab oil boycott in October 1973, although many other factors have contributed to the dilemma. This situation brought about an almost overnight recognition by the overall American public that energy sources are indeed valuable *resources*. Furthermore, the situation pointed out that many of these resources are in short supply, and that significant progress is essential in areas of conservation and development of domestic supplies, in order to meet projected demand requirements. For the future, energy/ environment questions will continue to be raised, and energy consideration in environmental impact analysis will take on an increasingly important role.

Although not specifically stated, implications for the inclusion of energy considerations in environmental impact analysis may be found in several sections of Title I of the National Environmental Policy Act. Recognizing that energy and fuels constitute a resource, perhaps the most obvious reference is made in Section 102(2)(C), where it is required that a detailed statement be made for federal actions on ". . . any irreversible and irretrievable commitments of resources which would be involved in the proposed action should it be implemented." Indirect implications are made in Section 101(b)(6), where it is stated that the federal government has the continuing responsibility to ". . . enhance the quality of renewable resources and approach the maximum attainable recycling of depletable resources."

These sections suggest at least four areas where energy considerations become a part of environmental impact analysis. These areas are (1) commitments of energy as a resource, (2) environmental effects of fuel resource development, (3) energy costs of pollution control, and (4) energy aspects of materials recycling. The following sections examine these particular areas relating energy considerations to the analysis of environmental impact.

Energy as a Resource

Energy resources include all basic fuel supplies that are utilized for heating, electrical production, transportation, and other forms of energy requirements. These resources may take the form of fossil fuels (oil, coal, gas, etc.), radioactive materials used in nuclear power plants, or miscellaneous fuels, such as wood, solid waste, or other combustible materials. Solar and hydroelectric energy resources or other energy sources currently in a developmental state may be significant for particular projects.

When a project consumes energy, this consumption should be considered as a primary or direct impact on resource consumption. Actions requiring consumption of energy can be categorized into (1) residential, (2) commercial, (3) industrial, and (4) transportation activities.

Residential activities include space heating, water heating, cooking, clothes drying, refrigeration, and air conditioning associated with the operation of housing facilities. Also included is the operation of energy-intensive appliances such as hair dryers, television sets, and the like.

Commercial activities include space heating, water heating, cooking, refrigeration, air conditioning, feedstock, and other energy-consuming aspects of facility operation. Facilities that consume particularly significant amounts of energy include bakeries, laundries, and hospital services.

Industrial activities that require large amounts of fuel resources include power plants, boiler and heating plants, and cold storage and air conditioning plants. Other industrial operations that require process steam, electric dryers, electrolytic processes, direct heat, or feedstock may impact heavily upon fuel resources.

Transportation activities involving the movement of equipment, materials, or persons require the consumption of fuel resources. The mode of transportation includes aircraft, automobile, bus, truck, pipeline, and watercraft.

The most important variables to be considered in determining impacts on fuel resources are the rate of fuel consumption for the particular activity being considered, and the useful energy output derived from the fuel being consumed. Various units may be utilized in describing consumption rates; miles per gallon, cubic feet per minute, and tons per day are commonly used in describing the consumption of gasoline, natural gas, and coal, respectively. Similarly, the energy output of various fuel-and-energy-consuming equipment and facilities may be described in many different units—horsepower, kilowatt-hours, and tons of cooling are a few examples.

A common unit of heat, the BTU, may be applied to most cases involving fuel or energy consumption. A BTU is the quantity of heat required to raise the temperature of one pound of water one Fahrenheit degree. In the evaluation of transportation systems, for example, alternatives may be compared on a BTU per passenger-mile or a BTU per ton-mile basis.

Other variables of concern include the availability (short-term and long-term) of fuel alternatives, cost factors involved, and transportation distribution and storage system features required for each alternative.

Data on the consumption of fuel resources may be applied to almost any environmental impact analysis, but the depth and degree to which such data is required depends upon the nature of the project under consideration. For an analysis of existing facilities or operations, sufficient information should be available from existing records and reference sources. Where alternative fuels or transportation systems are under consideration, additional background information may be necessary to evaluate not only efficiencies, but cost-effectiveness and long-term reliability.

Because of the complexities in the nature of the variables discussed above, most are measured by engineers or energy economists, although the results may be applied by most individuals with technical training.

Once the heat contents of fuels are known, comparisons may be made on the basis of the heat content of each required to achieve a given performance. An energy ratio (ER) can be established as the tool for comparison. The ER is defined as the number of BTU's of one fuel equivalent to one BTU of another fuel supplying the same amount of useful heat:

$$ER = \frac{\text{Amount of fuel No. 1 used} \times \text{heat content of fuel No. 1}}{\text{Amount of fuel No. 2 used} \times \text{heat content of fuel No. 2}}.$$

Determination of energy ratios requires careful testing in laboratory or field comparisons, but yields usually reliable results when conducted under impartial and competently supervised conditions. These ratios have been determined in various tests and are summarized in such publications as the *Gas Engineer's Handbook*.[16]

The consumption of fuels for a particular use may be determined from procurement and operational records. Measurements may be made using conventional meters, gauges, and other devices.

The fuel resource data can be used in an environmental impact analysis for the benefit of planners and decision makers for either (1) evaluating the alternatives where either fuel consumption or fuel-consuming equipment or facilities are involved; or (2) determining baseline fuel and energy consumption. This analysis includes the evaluation of irreversible and irretrievable commitments of resources resulting from the action, the short-term/long-term tradeoffs, and the identification of areas for potential conservation and mitigation of unnecessary waste. The analysis would include evaluation of efficiency, availability, cost of fuel and support facilities (transportation, distribution, storage, etc.), and projected changes in these values that might occur in the future.

Conversion of fossil and nuclear fuels into usable energy can lead to both direct and secondary effects on the biophysical and socio economic environment. Some of the effects that may occur are listed in Table 8.4. These impacts would also be considered in the analysis.

If a project results in significant additional demands or waste of fuels already in short supply, public controversy may be expected to follow. Natural gas supplies, presently limited or unavailable in some areas, should be considered with special emphasis. Electric consumption, in

TABLE 8.4 Environmental Effects Related
to Energy Consumption

Environmental Area	Environmental Problems
Air	Pollutant Emissions —Carbon monoxide —Sulfur oxide —Hydrocarbons —Nitrogen oxide —Lead —Mercury —Other toxic compounds —Smoke —Smog
Water	Oil spills Brines Acid mine drainage Heat discharges
Land	Land disturbance Aesthetic blight Loss of habitat Subsidence
Solid waste	Leachates Radioactive waste Storage/disposal waste

most cases, bears directly upon fuel resources, the effects of which should
be included in the analysis.

Concern for fuel resources typically peaks during summer (when air
conditioning loads are high) and winter (when heating loads are high).
Thus, projects in northern climates would be expected to have the greatest concern for heating fuels, while in the south, the emphasis would be
on projects with heavy cooling requirements in the summer, although ex-

**TABLE 8.5   Comparative Environmental Impacts of 1,000-Megawatt
Electric Energy Systems Operating at a 0.75 Load Factor
With Low Levels of Environmental Controls or With
Generally Prevailing Controls**

System	Air emissions Severity	Water discharges Severity	Solid waste Severity	Land use Severity	Potential for large-scale disaster
Coal					
Deep-mined	5	5	3	3	Sudden subsidence in urban areas, mine accidents
Surface-mined	5	5	5	5	Landslides
Oil					
Onshore	3	3	1	2	Massive spill on land from blowout or pipe-line rupture
Offshore	3	4	1	1	Massive spill on land from blowout or pipe-line rupture
Imports	2	4	1	1	Massive oil spill from tanker accident
Natural gas	1	2	0	2	Pipeline explosion
Nuclear	1	3	4	2	Core meltdown, radiological health accidents

Severity rating key:
5 = serious, 4 = significant, 3 = moderate, 2 = small, 1 = negligible, 0 = none.
SOURCE: U.S. Council on Environmental Quality. *Energy and the Environment—Electric
Power*, August, 1973.

ceptions to this may occur due to localized demands or geographical or climatic effects. Proximity to natural supplies also plays an important role in fuel selection, since transportation may affect the availability and economic desirability of certain fuels.

Mitigation of impacts directly and indirectly attributable to energy and fuel resources falls into two categories. The first pertains to mitigation by alternate fuel selection, and is based on a number of complex variables— availabilty, cost, environmental effects, and pollution control requirements, to name a few. Other factors to be considered in the selection are the short-term/long-term effects of a particular choice, and the irreversible and irretrievable commitment of resources associated with the selection. The second category of mitigation is associated with the conservation of energy, regardless of the type or types of fuel being consumed. These mitigations, however, bring up other environmental questions, as shown in the following sections.

Fuel Alternatives and Development of Supplies—Environmental Considerations

Not all fuel alternatives produce the same effects on the environment. The Council on Environmental Quality (CEQ) recently undertook a study of the environmental impact of electric power alternatives and concluded that such a comparative discussion is useful for discussion purposes and provides a basis for further analysis.[17] Table 8.5 presents a summary of the study that examined seven systems operating with low level or contemporary environmental controls. CEQ recognized the difficulty in making comparisons of very different systems and stressed that regional differences, emission control variability, and other factors should be considered in addition to the aggregated information presented. Next, each fuel is examined for specific environmental effects.

Coal. Although coal is our most abundant fossil fuel resource, its use in the production of electrical energy is judged the most environmentally damaging of alternatives. The following details some of the problems associated with the use of coal:

Operation	*Major Environmental Effects*
Surface mining	Land disturbance
	Acid mine drainage
	Silt production
	Solid waste
	Habitat disruption
	Aesthetic impacts

Operation	*Major Environmental Effects*
Underground mining	Acid drainage
	Land subsidence
	Occupational health & safety
	Solid waste
Processing	Solid waste stockpiles
	Waste water
Transportation	Land use
	Accidents
	Fuel utilization
Conversion	Air pollution
	Sulfur oxides
	Nitrogen oxides
	Particulates
	Solid wastes
	Thermal discharge
Transmission lines	Land use
	Aesthetics

Oil. Environmental effects from oil are different from those of coal, as may be seen from the following summary:

Operation	*Major Environmental Effects*
Extraction	Land use (drilling)
	Spillage
	Brine disposal
	Blowouts
Transportation	Land use (pipeline)
	Leakage and rupture (pipelines)
	Spills
Refining	Air pollution
	Water pollution
Conversion	Air pollution
	Sulfur oxides
	Nitrogen oxides
	Hydrocarbons
	Thermal discharge
Transmission lines	Land use
	Aesthetics

Gas.   Natural gas is significantly more desirable from a pollution production standpoint, as may be seen from the following table:

Operation	Major Environmental Effects
Extraction	Land use (drilling)
	Brine disposal
Transportation	Land use (pipeline)
Processing	Air pollution (minor)
Conversion	Air pollution (relatively minor)
	Carbon monoxide
	Nitrogen oxide
	Thermal discharge
Transmission lines	Land use
	Aesthetics

Nuclear Fission.   A different set of environmental effects results from the nuclear fission process, as indicated below.   The accident potential in conversion and disposal represents a highly controversial issue in evaluating nuclear fission processes.

Operation	Major Environmental Effects
Mining	Land use (not extensive)
Milling (separation)	Radioactive wastes
	Air
	Water
	Solid waste
Enrichment	Minor release of radioactive material
Conversion	Thermal discharge
	Release of radionuclides (minor)
	Accident potential
Transmission lines	Land use
	Aesthetics
Reprocessing	Radioactive air emissions
Radioactive waste disposal	Accident potential (handling, storing)

Fuel selection must be made on the basis of many factors in addition to environmental consequences—cost, availability, and facilities and equipment requirements also must be considered. Both short-term and long-term aspects would be included in the life-cycle analysis of a proposed system, and decision makers should consider all aspects in making fuel selections.

To mitigate or reduce the adverse environmental effects of energy and fuel utilization, various procedures have been initiated, some of which have been controversial, and/or have not been effective, and/or have resulted in further limitations in fuel supplies. For example, some fuel supplies have been restricted or limited through strip mine regulations and limitations on drilling and exploration.

The conservation of energy may be accomplished through (1) voluntary means, such as cutbacks in heating and lighting use, (2) economic incentives, such as taxation, or (3) legislative means, such as mandatory speed limits. Conservation will undoubtedly continue to play a key role as the nation moves toward energy self-sufficiency.

Conservation measures may vary greatly with project type and magnitude. Such measures can be applied to new construction, in the form of additional insulation and design, incorporating energy conservation features related to color, orientation, shape, lighting, etc. Conservation of energy can be applied to existing facilities, in the form of added insulation and programs to reduce loads on heating, cooling, and other utility consumption. Likewise, in the operation and maintenance of equipment, steps may be taken to reduce fuel consumption further by increasing efficiencies through proper equipment maintenance, reducing transportation requirements, and scheduling replacement of old equipment with newer, highly efficient models.

Special efforts toward energy conservation should be pointed out in an environmental impact statement because, generally, adverse impacts on the biophysical environment tend to become reduced with decreased energy production and consumption. However, some question arises as to socioeconomic effects of a substantially lowered growth rate of energy consumption.

Energy Costs of Pollution Control

Energy requirements for the operation of pollution control systems are an area of conflict that probably will continue to be present as long as pollution regulations are in effect. One study has shown that the quantities of energy required to achieve various environmental quality goals are relatively small—3.5 to 4.0 percent of the U.S. total energy use.[18] However, overall increases in energy costs as a result of pollution controls

TABLE 8.6   Expected Cost of Environmental Controls

Fuel Source	% Total Cost Increase for Pollution Control
Coal	28–31
Oil	34–36
Gas	5
Nuclear	5

SOURCE: U.S. Council on Environmental Quality. *Energy and the Environment—Electric Power*, USGPO, August 1973.

will occur, as shown in Table 8.6. These increases represent modifications to existing energy systems by adding additional controls (present technology; cooling towers, land reclamation, air pollution controls, etc.) to meet current standards.

Generally speaking, the energy requirements for various aspects of pollution control will vary with type of process, quantities involved, and degree of treatment or removal. The following discussion addresses the major pollution control features in current use.

Wastewater Treatment.   Electric requirements for sewage treatment can be related directly to plant size, as shown in Fig. 8.3. Since the per capita energy requirements diminish with increasing plant size, it would appear that consolidation of waste treatment facilities with surrounding communities would be advantageous from an energy conservation standpoint.

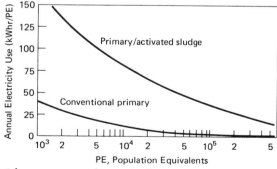

Fig. 8-3.   Electric energy requirements for sewage treatment as a function of plant size.

SOURCE: "Energy Implications of Cleanup Operations," by Eric Hirst, *Environmental Science and Technology*, January 1975.

TABLE 8.7   Energy Requirements for Air Pollution Controls

Pollutant	Process	Energy Requirement (Approx.)
Particulates	Electrostatic precipitators	0.1% plant output
Sulfur oxides	Limestone scrubbing	4–8% plant output
Sulfur oxides	Molten carbonate	1% plant output

SOURCE: "Energy Implications of Cleanup Operations," by Eric Hirst, *Environmental Science and Technology*, January 1975.

Solid Waste Management. Energy use for solid waste collection, transport, and landfilling has been estimated at approximately 300,000 BTU/ton.[18] On the other hand, incineration of solid waste refuse can yield heat at about 10 million BTU/ton, and can be used to generate electricity. Air pollution, health, and other problems may occur, however.

Air Polution Control. Energy consumption by air pollution controls on stationary power sources is relatively small, as indicated in Table 8.7.

Other air pollution control measures (transportation, industrial processes, etc.) are frequently cited as the basis for increased production costs and decreased operating efficiencies. However, it has been shown that adequate control equipment can operate effectively with relatively small additional operating expense.

After examining the energy aspects of pollution control, several conclusions may be drawn:

1. Present total energy growth rates cannot be attributed to operation of environmental protection systems.
2. Energy needs for environmental protection may be offset by a few simple conservation measures (recycling, insulating, design, etc.).
3. Claims that higher costs and shortages in supplies are due to *energy* requirements for pollution controls are questionable.

Energy Aspects of Materials Recycle

The recycling of materials such as paper, metals, and glass has long been known to reduce environmental problems such as solid waste, litter, etc., while at the same time conserving supplies and preserving resources. Recently, however, a renewed look at recycling has come about as a result of the energy aspect of materials manufacture.

Some indication of the potential for energy conservation may be revealed upon an examination of the energy requirements for various sectors as indicated in Table 8.8, and the distribution of energy consumption in the manufacturing sector shown in Table 8.9. The primary products

TABLE 8.8   Distribution of Energy Consumption by Sector: 1971

	Purchased Fuels		Purchased Fuels Plus Electricity*	
	BTU	%	BTU	%
Household/commercial	14,281	20.7	17,441	30.6
Transportation	16,971	24.6	16,989	29.8
Industrial	20,294	29.4	22,623	39.7
Manufacturing	(14,329)	(20.8)	(16,085)	(27.9)
Non-manufacturing	(5,965)	(8.6)	(6,538)	(11.7)
Electrical generation	17,443	25.3	—	—
TOTAL	68,989	100	57,053	100

*Purchased electricity valued at its thermal equivalence of 3,412 Btu/kwh.
SOURCE: Federal Energy Administration. "Energy Conservation in the Manufacturing Sector 1954–1980," *Project Independence Blueprint Final Task Force Report* 3, November 1974.

TABLE 8.9   Distribution of Energy Consumption Within the Manufacturing Sector: 1971

Six Energy Intensive Manufacturing Industries	Purchased Fuels		Purchased Fuels and Electricity	
	Btu	%	Btu	%
(1) Food and kindred products	800	5.6	920	5.7
(2) Paper and allied products	1196	8.3	1315	8.2
(3) Chemicals and allied products	2443	17.0	2783	17.3
(4) Petroleum and coal products	2877	20.0	2956	18.4
(5) Stone, clay and glass products	1291	9.0	1367	8.5
(6) Primary metals industries	3613	25.2	4030	25.1
Total of six	12220	85.1	13371	83.2
Other Manufacturing	2109	14.7	2714	16.9
Total manufacturing	14329	99.8[a]	16085	100.1[a]

[a]Failure to sum to 100% due to rounding error.
SOURCE: Federal Energy Administration. "Energy Conservation in the Manufacturing Sector 1954–1980," *Project Independence Blueprint Final Task Force Report* 3, November 1974.

industries (Food, Paper, Chemical, Petroleum, Stone, Clay and Glass, and Primary Metals) in 1971 accounted for over 83 percent of the energy consumed by manufacturing, or 23.2 percent of the total U.S. requirements for that year. As energy prices escalate and uncertainty surrounds future supplies, these industries are forced to examine programs to improve their energy efficiencies.[19] One approach that is advocated is recycling.

Recycling and recovery of materials from waste streams depends primarily on economics. The cost of manufacturing products from recycled materials has been essentially the same or, in some cases, even higher than that of manufacturing the products from raw materials.[20] Depletion allowances, capital gains treatments, transportation costs, and other factors have essentially inhibited a greater movement toward recovery efforts. Recent increases in energy costs along with increases in material costs and shortages in many materials have stimulated recycling through the creation of new markets and demands for recycled products.[21]

Recycle of Specific Materials

The following discussion examines several recyclable products in view of energy and other environmental considerations.

Glass. Glass can be recycled back into glass furnaces, but difficulties in the glass-making operation present problems that make recycling unattractive. First, glass "formulas" include not only silica, but limestone, soda ash, and, in many cases, coloring agents that are blended, melted, and refined in precise operations. Reclaimed glass results in the blending of formulas and many foreign substances, the end products of which are highly unpredictable.

The separation of glass from other wastes poses a second problem to glass recycling. This process may vary from simple hand-classification, accomplished during time of collection, to complex automated separation operations employing air classification, dense media separation, or froth flotation. Color separation must also be accomplished and may be done at time of collection or via automated optical systems.[22]

Finally, energy and transportation costs are such that the economics of manufacturing glass from ore is essentially the same as that from recycled materials.[19] Thus, glass recycling, at the present, is limited to factories near collection centers and special cases in which the process can be justified economically.

Utilization of returnable bottles and containers assures that the effective use of a given container will be greatly increased, thereby decreasing the necessity for more containers and the waste produced as each container is emptied. Discouragement of "throw-away" containers not only promotes less waste production, but less energy expenditure for manufacturing as well. When the total energy consumption involved in collecting, returning, washing, and refilling glass bottles is compared to that required in delivering the same volume of beverage to the consumer in a "throw-away" container, a significant energy savings is apparent. One study has indicated that "... a complete conversion to returnable bottles would reduce the demand for energy in the beverage (beer and soft drink) industry by 55 percent, without raising the price of soft drinks to the consumer."[23]

Reclaimed glass may be used for secondary products other than glass containers, such as for aggregate in road construction, manufacture of insulating materials, or in brick production.[22]

Tires and Rubber Products.   Rubber is a natural forest products resource that is critical to military and civilian transportation and to the production of mechanical rubber goods.   Natural rubber is used primarily for tire production, and approximately 80 percent of the U.S. requirements come from Indonesia and Malaysia.   Disposal of tires and other rubber products represents a potential loss in several ways.   Disposal represents a problem from an economic standpoint of collection, shipment, storage, and ultimate disposal.   Disposal of rubber goods presents an environmental question, as the long term effects of slowly disintegrating tires and rubber products have not been determined.

Recycle and reuse potential for scrap tires and rubber products include (1) direct reuse as artificial reef construction, (2) reprocessing for retreaded tires or other rubber products, (3) alternate use such as in road surfacing, or (4) as a fuel in boilers.   All these represent a possible resource enhancement or savings, and some are directly or indirectly related to energy savings as well.

Paper.   Recycled paper can be manufactured relatively easily, with end products competitive in quality to those made from virgin materials.   Major difficulties arise from the economics of collection and transportation of waste paper products to centers for reprocessing.   Significant environmental and energy conservation features result from direct recycling— a 70 percent reduction in energy requirements may be possible in the manufacture of low-grade paper.[20]

Shredded waste paper and other forms of waste paper products may be utilized as packaging materials, mulches for erosion control, or may form a portion of compost material for soil enrichment.   When solid waste is utilized for incineraton and heat recover, paper content is converted to heat for energy production.

Disposal problems associated with paper waste include collection and transportation cost and difficulties normally associated with solid waste management.   Inasmuch as paper is an organic material, decomposition in landfills produces gas and leachate hazard potential.   Chemicals contained in inks and dyes create additional problems in disposal and in use for compost materials.

Metals. High costs of metals and metal products have resulted in extensive programs to reclaim many—stainless steel, precious metals, lead, and copper, in particular. Significant amounts of steel, aluminum, and zinc are also recycled, but not to the extent that could, or perhaps should, be returned for reuse. As with other waste materials, metal recycling reduces the quantity of solid waste to be disposed, reduces the consumption of natural resources, and further reduces the energy requirements for the production of manufactured products.

In steel production, for example, 74 percent less energy is required to produce 1,000 tons of steel reinforcing bars from scrap than from ore.[20] Similar savings are attained in the production of aluminum and other metals from scrap.

Oil Wastes. Waste oil and petroleum products originate from crankcase and lubrication wastes generated during the normal maintenance of motorized vehicles and machinery. Waste oils may be used directly without reprocessing as road oils for dust control, or may be mixed with virgin fuel oil for use in boilers for heating or electrical power generation.

The process of refining waste oil to produce lubrication oils or fuel oils is technologically possible and currently is being practiced in many areas. Difficulties in removing impurities of lead, dirt, metals, oxidation products, and water, along with environmental standards and product specifications, have hampered the widespread practice of recycling in the past. However, the shortage of petroleum, coupled with economic incentives, may result in a resurgence in recycling of petroleum products in the near future.

Waste oil and its impurities possess potential threats to the environment, whether the waste oil is indiscriminately dumped on land or into water courses or burned. Even the refining process may produce acid sludges and contaminated clays that must be disposed of in a manner that is environmentally safe.

General Solid Waste. Municipal solid waste has been termed by some an "urban ore" with a great potential for materials and energy recovery. Currently, a great variety of approaches are being investigated and demonstrated to tap this potential resource. Typical content includes the following:

Paper	Lead
Glass	Textiles
Ferrous metals	Rubber
Aluminum	Plastics
Tin	Food, animal, plant and other wastes
Copper	Miscellaneous materials

In addition to materials recovery and the potential savings represented, many solid wastes may be incinerated with significant energy recovery. The theoretical energy value available by 1980 through energy generation from municipal solid waste has been estimated at approximately equivalent to one million barrels of oil per day.[21]

Summary

It is now a well-documented and generally accepted fact that our environment has suffered greatly at the expense of the development of our present life styles. It is also now being recognized that we cannot maintain these life styles if fuels and other non-renewable resources continue to be consumed at rates that indicate depletion of many critical materials in a matter of only a few decades—in some cases, only a few years. Demands for a higher-quality environment coupled with shortages in fuels and other commonplace items have shown that hard decisions regarding the trade-offs between energy consumption, pollution control, and life styles will have to be made to insure long-term viability of the planet Earth as a place for man to live.

REFERENCES

1. Sax, Joseph, *Defending the Environment: A Handbook for Citizen Action*, 58, 1970.

2. Manty, Dale, *et al.*, "Conceptual Framework for Effective Community Participation," *A Report to the Battelle Urban and Regional Development Program*, 1974.

3. Ayres, R. V., and A. V. Kneese, "Production Consumption, and Externalities," *Am. Eco. Rev.* **59**: 3–14 (1970).

4. Isard, W., *Ecologic-Economic Analysis for Regional Development*, The Free Press, New York, 1972.

5. Cumberland, J. H., *Regional Development Experience in the United States of America*, 170, The Hague, Mouton, 1971.

6. Leontief, W., "Environmental Repercussions and the Economic Structure: An Input-Outut Approach," *A Challenge to Social Scientists*, Shiegeto Tsuru (ed.), Asahi, Tokyo, 114–134, 1970.

7. Laurent, E. A. and J. C. Hite, *Economic-Ecologic Analysis in the Charleston Metropolitan Region: An Input-Output Study*, Water Resources Research Institute, Clemson University, April 1971.

8. Davis, R. M., G. S. Stacey, G. I. Nehman, and F. K. Goodman, "Development of an Economic-Environmental Trade-Off Model for Industrial Land Use Planning," *Rev. Regional Stud.* **4**: 1, Spring 1974.

9. Ide, E. A., *Estimating Land and Floor Area Implicit in Employment—How Land and Floor Area Usage Rate Vary by Industry and Site Factors*, U.S. Department of Transportation, July 1970.

10. *Ibid.*, 1–2.

11. Laurent, E. A. and J. C. Hite, *Op cit.*

12. Stepp, J. M., *Water Use, Waste Treatment, Water Pollution and Related Economic Data on South Carolina Manufacturing Plants*, Clemson University, Water Resources Institute, *No. 8*, 1968.

13. Duprey, J. M., "Compilation of Air Pollutant Emission Factors," HEW National Air Pollution Control Administration, Durham, *999-AP-42*, 1968.

14. Webster, R. D., *Development of the Environmental Technical Information System*, IR E-52, U.S. Army Construction Engineering Research Laboratory, Champaign, Illinois.

15. Miernyk, W. H., "Long Range Forecasting With a Regional Input-Output Model," *Western Econ. J.* **VI**, *No. 3* (June 1968).

16. *Gas Engineers Handbook*, Chapter 22, "Fuel Comparisons," The Industrial Press, New York, 1965.

17. Council on Environmental Quality (CEQ), *Energy and the Environment—Electric Power*, USGPO, August 1973.

18. Hirst, E., "Energy Implications of Cleanup Operations," *Environmental Sci. and Tech.* **9** *No. 1* (January 1965).

19. Federal Energy Administration, "Energy Conservation in the Manufacturing Sector 1954–1989," *Project Independence Blueprint Final Task Force Report*, 3 USGPO, November 1974.

20. Environmental Protection Agency, *First Report to Congress—Resource Recovery and Source Reduction*, (Delivered February 22, 1073), SW-118, (3rd Ed., 1974).

21. CEQ, *The Fifth Annual Report of the Council on Environmental Quality*, USGPO, December 1974.

22. "Glass Recycling Makes Strides," *Environmental Sci. and Tech.* **6**, *No. 12* (November 1972).

23. Hannon, B. M., "Bottles, Cans, Energy," *Environment* **14**, *No. 2*, (March 1972).

24. *Environmental Quality*, Fourth Annual Report of the Council on Environmental Quality, USGPO, September 1973.

Appendix A

**THE NATIONAL ENVIRONMENTAL POLICY
ACT—PUBLIC LAW 91-190**
(As Amended by PL 94-83)

PURPOSE

Sec. 2. The purposes of this Act are: To declare a national policy which will encourage productive and enjoyable harmony between man and his environment; to promote efforts which will prevent or eliminate damage to the environment and biosphere and stimulate the health and welfare of man; to enrich the understanding of the ecological systems and natural resources important to the Nation; and to establish a Council on Environmental Quality.

TITLE I
DECLARATION OF NATIONAL
ENVIRONMENTAL POLICY

Sec. 101. (a) The Congress, recognizing the profound impact of man's activity on the interrelations of all components of the natural environment, particularly the profound influences of population growth, high-density urbanization, industrial expansion, resource exploitation, and new and expanding technological advances and recognizing further the critical importance of restoring and maintaining environmental quality to the overall welfare and development of man, declares that it is the continuing policy of the Federal Government, in cooperation with State and local governments, and other concerned public and private organizations, to use all practicable means and measures, including financial and technical assistance, in a manner calculated to foster and promote the general welfare, to create and maintain conditions under which man and nature can exist in productive harmony, and fulfill the social, economic, and other requirements of present and future generations of Americans.

(b) In order to carry out the policy set forth in this Act, it is the continuing responsibility of the Federal Government to use all practicable means, consistent with other essential considerations of national policy, to improve and coordinate Federal plans, functions, programs, and resources to the end that the Nation may—

(1) fulfill the responsibilities of each generation as trustee of the environment for succeeding generations;

(2) assure for all Americans safe, healthful, productive, and esthetically and culturally pleasing surroundings;

(3) attain the widest range of beneficial uses of the environment without degradation, risk to health or safety, or other undesirable and unintended consequences;

(4) preserve important historic, cultural, and natural aspects of our national heritage, and maintain, wherever possible, an environment which supports diversity and variety of individual choice;

(5) achieve a balance between population and resource use which will permit high standards of living and a wide sharing of life's amenities; and

(6) enhance the quality of renewable resources and approach the maximum attainable recycling of depletable resources.

(c) The Congress recognizes that each person should enjoy a healthful environment and that each person has a responsibility to contribute to the preservation and enhancement of the environment.

Sec. 102. The Congress authorizes and directs that, to the fullest extent possible: (1) the policies, regulations, and public laws of the United

States shall be interpreted and administrated in accordance with the policies set forth in this Act, and (2) all agencies of the Federal Government shall—

(A) utilize a systematic, interdisciplinary approach which will insure the integrated use of the natural and social sciences and the environmental design arts in planning and in decisionmaking which may have an impact on man's environment.

(B) identify and develop methods and procedures, in consultation with the Council on Environmental Quality established by title II of this Act, which will insure that presently unquantified environmental amenities and values may be given appropriate consideration in decisionmaking along with economic and technical considerations;

(C) include in every recommendation or report on proposals for legislation and other major Federal actions significantly affecting the quality of the human environment, a detailed statement by the responsible official on—

(i) the environmental impact of the proposed action,

(ii) any adverse environmental effects which cannot be avoided should the proposal be implemented,

(iii) alternatives to the proposed action,

(iv) the relationship between local short-term uses of man's environment and the maintenance and enhancement of long-term productivity, and

(v) any irreversible and irretrievable commitments of resources which would be involved in the proposed action should it be implemented.

Prior to making any detailed statement, the responsible Federal official shall consult with and obtain the comments of any Federal agency which has jurisdiction by law or special expertise with respect to any environmental impact involved. Copies of such statement and the comments and views of the appropriate Federal, State, and local agencies, which are authorized to develop and enforce environmental standards, shall be made available to the President, the Council on Environmental Quality and to the public as provided by section 552 of title 5, United States Code, and shall accompany the proposal through the existing agency review processes;

(D) Any detailed statement required under subparagraph (C) after January 1, 1970, for any major Federal action funded under a program of grants to States shall not be deemed to be legally insufficient solely by reason of having been prepared by a State agency or official, if:

(i) the State agency or official has statewide jurisdiction and has the responsibility for such action,

(ii) the responsible Federal official furnishes guidance and participates in such preparation.

(iii) the responsible Federal official independently evaluates such statement prior to its approval and adoption, and

(iv) after January 1, 1976, the responsible Federal official provides early notification to, and solicits the views of, any other State or any Federal land management entity of any action or any alternative thereto which may have significant impacts upon such State or affected Federal land management entity and, if there is any disagreement on such impacts, prepared a written assessment of such impacts and views for incorporation into such detailed statement.

The procedures in this subparagraph shall not relieve the Federal official of his responsibilities for the scope, objectivity, and content of the entire statement or of any other responsibility under this Act; and further, this subparagraph does not affect the legal sufficiency of statements prepared by State agencies with less than statewide jurisdiction.

(E) study, develop, and describe appropriate alternatives to recommended courses of action in any proposal which involves unresolved conflicts concerning alternative uses of available resources;

(F) recognize the worldwide and long-range character of environmental problems and, where consistent with the foreign policy of the United States, lend appropriate support to initiatives, resolutions, and programs designed to maximize international cooperation in anticipating and preventing a decline in the quality of mankind's world environment;

(G) make available to States, counties, municipalities, institutions, and individuals, advice and information useful in restoring, maintaining, and enhancing the quality of the environment;

(H) initiate and utilize ecological information in the planning and development of resource-oriented projects; and

(I) assist the Council on Environmental Quality established by title II of this Act.

Sec. 103. All agencies of the Federal Government shall review their present statutory authority, administrative regulations, and current policies and procedures for the purpose of determining whether there are any deficiencies or inconsistencies therein which prohibit full compliance with the purposes and provisions of this Act and shall propose to the President not later than July 1, 1971, such measures as may be necessary to bring their authority and policies into conformity with the intent, purposes, and procedures set forth in this act.

Sec. 104. Nothing in Section 102 or 103 shall in any way affect the specific statutory obligations of any Federal agency (1) to comply with criteria or standards of environmental quality, (2) to coordinate or consult with any other Federal or State agency, or (3) to act, or refrain from acting contingent upon the recommendations or certification of any other Federal or State agency.

Sec. 105. The policies and goals set forth in this Act are supplementary to those set forth in existing authorizations of Federal agencies.

TITLE II
COUNCIL ON ENVIRONMENTAL QUALITY

Sec. 201. The President shall transmit to the Congress annually beginning July 1, 1970, an Environmental Quality Report (hereinafter referred to as the "report") which shall set forth (1) the status and condition of the major natural, manmade, or altered environmental classes of the Nation, including, but not limited to, the air, the aquatic, including marine, estuarine, and fresh water, and the terrestrial environment, including, but not limited to, the forest dryland, wetland, range, urban, suburban, and rural environment; (2) current and foreseeable trends in the quality, management and utilization of such environments and the effects of those trends on the social, economic, and other requirements of the Nation; (3) the adequacy of available natural resources for fulfilling human and economic requirements of the Nation in the light of expected population pressures; (4) a review of the programs and activities (including regulatory activities) of the Federal Government, the State and local governments, and nongovernmental entities or individuals, with particular reference to their effect on the environment and on the conservation, development and utilization of natural resources; and (5) a program for remedying the deficiencies of existing programs and activities, together with recommendations for legislation.

Sec. 202. There is created in the Executive Office of the President a Council on Environmental Quality (hereinafter referred to as the "Council"). The Council shall be composed of three members who shall be appointed by the President to serve at his pleasure, by and with the advice and consent of the Senate. The President shall designate one of the members of the Council to serve as Chairman. Each member shall be a person who, as a result of his training, experience, and attainments, is exceptionally well qualified to analyze and interpret environmental trends and information of all kinds; to appraise programs and activities of the Federal Government in the light of the policy set forth in title I of this Act; to be conscious of and responsive to the scientific, economic, social, esthetic, and cultural needs and interests of the Nationa; and to formulate and recommend national policies to promote the improvement of the quality of the environment.

Sec. 203. The Council may employ such officers and employees as may be necessary to carry out its functions under this Act. In addition, the

Council may employ and fix the compensation of such experts and consultants as may be necessary for the carrying out of its functions under this Act, in accordance with section 3109 of title 5, United States Code (but without regard to the last sentence thereof).

Sec. 204. It shall be the duty and function of the Council—

(1) to assist and advise the President in the preparation of the Environmental Quality Report required by section 201;

(2) to gather timely and authoritative information concerning the conditions and trends in the quality of the environment both current and prospective, to analyze and interpret such information for the purpose of determining whether such conditions and trends are interfering, or are likely to interfere, with the achievement of the policy set forth in title I of this Act, and to compile and submit to the President studies relating to such conditions and trends;

(3) to review and appraise the various programs and activities of the Federal Government in the light of the policy set forth in title I of this Act for the purpose of determining the extent to which such programs and activities are contributing to the achievement of such policy, and to make recommendations to the President with respect thereto;

(4) to develop and recommend to the President national policies to foster and promote the improvement of environmental quality to meet the conservation, social, economic, health, and other requirements and goals of the Nation;

(5) to conduct investigations, studies, surveys, research, and analyses relating to ecological systems and environmental quality;

(6) to document and define changes in the natural environment, including the plant and animal systems, and to accumulate necessary data and other information for a continuing analysis of these changes or trends and an interpretation of their underlying causes;

(7) to report at least once each year to the President on the state and condition of the environment; and

(8) to make and furnish such studies, reports thereon, and recommendations with respect to matters of policy and legislation as the President may request.

Sec. 205. In exercising its powers, functions, and duties under this Act, the Council shall—

(1) consult with the Citizens' Advisory Committee on Environmental Quality established by Executive Order numbered 11472, dated May 29, 1969, and with such representatives of science, industry, agriculture, labor, conservation organizations, State and local governments, and other groups as it deems advisable; and

(2) utilize, to the fullest extent possible, the services, facilities, and information (including statistical information) of public and private

agencies and organizations, and individuals, in order that duplication of effort and expense may be avoided, thus assuring that the Council's activities will not unnecessarily overlap or conflict with similar activities authorized by law and performed by established agencies.

Sec. 205. Members of the Council shall serve full time and the Chairman of the Council shall be compensated at the rate provided for Level II of the Executive Schedule Pay Rates (5 U.S.C. 5313). The other members of the Council shall be compensated at the rate provided for Level IV of the Executive Schedule Pay Rates (5 U.S.C. 5315).

Sec. 207. There are authorized to be appropriated to carry out the provisions of this Act not to exceed $300,000 for fiscal year 1970, $700,000 for fiscal year 1971, and $1,000,000 for each fiscal year thereafter.

Appendix B

ATTRIBUTE DESCRIPTOR PACKAGE

It should be noted that these descriptions are intended to give the reader an overview of each attribute in the context of its role in impact analysis. None of the descriptions should be considered complete, as indeed, many of the individual subject areas themselves form the basis for complete texts. It is anticipated that familiarity with these 49 attributes can serve to expedite interdisciplinary studies which frequently encounter difficulties due to lack of communication between disciplines. This communication problem can be overcome when the participants attain some understanding of each other's terminology, problems, and difficulties in achieving solutions to those problems.

AIR

Air attributes are factors that indicate the quality of the air. Basically, two kinds of environmental factors relate to air quality. They relate to:

Structure elements of the environment
Inputs to or emissions from human activities.

Factors relating to the structure and elements of the air environment are stability, temperature, mixing depth, wind speed, wind direction, humidity, precipitation, pressure and topography. On the other hand, factors relating to inputs from human activity are dust, fumes, flyash, smoke, soot, and compounds of arsenic, aluminum, etc.

Nine attributes may be utilized in describing the impact of human activities on the air environment:

Diffusion factor
Particulates
Sulfur oxides
Hydrocarbons
Nitrogen oxides
Carbon monoxide
Photochemical oxidants
Hazardous toxicants
Odors.

The first attribute, the diffusion factor, is related to the structure and elements of the environment; the remaining attributes are related to the emissions from human activities.

Diffusion Factor

Definition of the Attribute. Diffusion factor is an attribute that is related to various atmospheric and topographic attributes of the environment. For example, vertical temperature structure affects movement of air in the atmosphere. Wind structure in a region determines the scavenging action in the environment as well as the impact of inversions. Topography may change temperature and wind profiles because of the combined effects of surface friction, radiation, and drainage. Valleys are more susceptible to stagnation and to air pollution than are flat lands or hill slopes. The mixing depth, in fact, also determines the intensity of air pollution in a given region. The status of stability or instability of the atmosphere determines to what extent air pollution can build up in a given region. Humidity and pressure also affect the diffusion rate of a given pollutant emitted to the atmosphere. In addition, precipitation is an important scavenger element that can clean up pollutants in the air.

Together, all of the above environmental factors determine the diffusion factor in a given region.

Activities that Affect the Attribute. Generally, most human activities will not affect the diffusion factor. However, since research has shown the possibility of certain activities affecting the weather and other related meteorological factors, it is necessary to consider such activities that are now known (about which limited information is available) which may impact the diffusion factor. For instance, artificial methods for generating storms, seeding clouds, and research and testing of these new and powerful methods can, and will, cause changes in the diffusion factor.

Source of Effects. As indicated above, impacts of certain specialized activities can have a major effect on the diffusion factor. Weather modification, in terms of cloud seeding, hail suppression, or alternate forms may affect precipitation patterns and other atmospheric attributes. The effects must be examined on a case-by-case basis, and where details of these activities are classified, for security reasons, it is not possible to provide detailed information on their potential impacts.

Variables to be Measured. Variables to be measured to determine the diffusion factor are many. The major ones are stability, mixing depth, wind speed, precipitation, and topography. Various measures of each of these variables will indicate the extent and nature of the diffusion factor in a given region.

How Variables are Measured. Generally, data on stability, mixing depth, wind speed, direction, and precipitation are collected by meteorological survey stations of the National Weather Service. Data on these attributes are readily available from the Weather Service offices across the country.

Topography data can be obtained from the United States Geological Survey (USGS) maps of largest available scale.

Data sources. Primary sources of data for the variables that define diffusion factors are the National Weather Service and the USGS; both have offices in most major cities throughout the country.

Skills required. Collection and analysis of such data require a sophisticated meteorological background. Persons with specialized training in meteorology and trained technicians are required to collect and develop information relating to these variables.

Instruments. A full-scale meteorological laboratory is needed to monitor the selected attributes that define the nature and extent of the air diffusion factor.

Evaluation and Interpretation of Data. The diffusion factor can be classified into three or more major ratings. For example, the diffusion factor can be High, Medium, or Low. The High rating represents an environmental quality (EQ) value of 1.0; the Medium rating represents an EQ value of 0.5; and the Low (or poor) rating represents an EQ value of 0.

The environmental impact of selected activities on diffusion factor is measured by the change in diffusion factor ratings. When a diffusion factor changes only a small amount and its rating remains unaltered, the impact is considered Insignificant. When the change in the diffusion factor rating is altered by one step (e.g., between High and Medium or Medium and Poor) the impact is considered to be Moderate. When a change in the diffusion factor rating occurs through two steps (e.g., between High and Poor) the impact is treated as Significant.

Geographical and Temporal Limitations. There can be substantial variation in the diffusion factor, spatially and temporally, depending upon variations in the determinant variables. It is known, for instance, that wind speed, precipitation, stability, and mixing depth change with time and location in a given region. These variations, therefore, alter the diffusion factor accordingly.

Mitigation of Impact. Generally, the impact of most activities on diffusion has not been adequately defined. The mitigation techniques are also not well established.

Secondary Effects. The diffusion factor can be related to land use patterns in the vicinity of air pollution sources. Prevailing wind direction can render certain land areas aesthetically undesirable or otherwise environmentally unacceptable during specific times or seasons of the year, if the air pollution is not eliminated or satisfactorily dispersed.

Other Comments. Research is needed to identify potential activities, their impacts, and the mitigation strategies relating to potential impacts on the diffusion factor. Also, a mathematical model is needed to relate all of the determinant variables to the diffusion factor. This will help establish a suitable relationship between variables and the diffusion factor.

Additional References. Stern, A., *Air Pollution* **I**, pp. 80–117, Academic Press, New York, 1962.

Particulates

Definition of the Attribute. Particulates are the most prevalent air pollutant. They exist in the form of minute separate particles suspended in the air.

Particulates are finely divided solid and liquid particles suspended in the ambient air. They range from over 100 microns to less than 0.01 microns in diameter. Particulates of smaller size (less than 10 microns) suspended in air can scatter light and behave like a gas. These smaller particulates are called aerosols.

Activities that Affect the Attribute. Many human activities generate particulates that are emitted to the air. These include construction, operation, maintenance, and repair activities; transportation; and industrial activities. Examples of subactivities are site preparation, demolition, removal and disposal; excavation; concrete construction; operation and maintenance of aircraft; operation and maintenance of automotive equipment; use of construction equipment; use of explosives; mineral extraction; foundry operations; manufacturing; noninitiating high explosives; and use of transportation vehicles.

Souirce of Effects. In general, the environment contains a certain level of particulate matter. Emissions, resulting from various activities, are released to the environment, causing a higher concentration of particulates. Particulates can cause increased mortality and morbidity in the exposed population by aggravating diseases such as bronchitis, emphysema, and cardiovascular diseases. Particulates can soil clothes and buildings and can cause serious visibility problems. Steel and other metal structures can be corroded as a result of exposure to particulates and humidity. Property values and psychic welfare of people can be undermined.

Variables to be Measured. Particulate concentration is generally measured as the concentration of all solid and liquid particles averaged over a period of 24 hours. For purposes of impact assessment, particulate con-

B

centration is measured as the average annual arithmetic mean of all 24 hour particulate concentrations at a given location.

How Variables are Measured. Particulate concentratons are usually measured by a High Volume Method, as specified in the *Federal Register* of April 30, 1971.[2] The air is drawn into a covered housing, through a filter, by a high-low blower at a rate of 40–60 cubic feet per minute. The particles, ranging from 100 to 0.1 microns in diameter, are ordinarily collected on fiberglass filters. The concentration of suspended particulates is then computed by measuring the mass of collected particulates in the volume sample in micrograms per cubic meter.

Data source. Sources of data are generally state pollution control departments, county air pollution control offices, multicounty air pollution control offices, or city air pollution control offices. High volume samplers may be installed to monitor particulates from specific operations.

Skills required. Basic paraprofessional training in mechanical or chemical engineering with special training in operating high volume samplers is adequate to collect particulate concentration data. Specialized supervision is needed to ensure that data are properly collected and analyzed.

Instruments. The apparatus used for sampling particulate concentration is called a High Volume Air Sampler. The sampler is installed in a shelter to protect it against extremes of temperature, humidity, and other weather conditions. It has a filter medium with a collection efficiency of about 99 percent for particles of 0.3 microns in diameter.

Evaluation and Interpretation of Data. The primary effects of particulates on environmental quality range from visibility problems to health impairments. Visibility problems occur at concentrations as low as 25 $\mu g/m^3$. As the concentration of particulates increases to about 200 $\mu g/m^3$, human health begins to be affected. The concentration levels mentioned above refer to 24 hour average annual concentration. Particulate concentration of less than 25 $\mu g/m^3$ is also considered less desirable for the environment, since it provides condensation nuclei upon which fog and cloud droplets settle. From these considerations, a particulate value function was developed, based on a 24 hour average annual concentration, as shown in Fig. B-1.

The determination of environmental impact of proposed activities on particulate level is measured by the change in particulates concentration. When the particulates concentration changes to the extent that its rating remains unaltered (e.g., High Quality air remains High Quality), the impact is considered Insignificant. When a change in particulates rating occurs through two steps (e.g., between High Quality to Low Quality, and vice versa), the impact is treated as Significant; a one step change is considered Moderate.

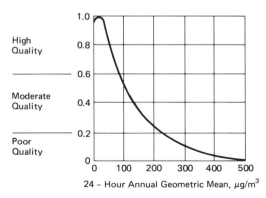

Fig. B-1 These value functions are provided for conceptual evaluation of air-quality impact. It should be noted that any time air-quality standards established by the governing regulatory agency are exceeded, the impact is significant.

The particulate value function (Fig. B-1) is used for rating air quality in terms of High, Moderate, and Low Quality, based on 24 hour annual geometric mean. For a given value of 24 hour annual geometric mean particulates concentration on the horizontal axis, a point on the curve identifies the environmental quality rating from the vertical axis of Fig. B-1 (e.g., 130 $\mu g/m^3$ indicates a moderate quality of 0.4).

Geographical and Temporal Limitations. Concentration of particulates does not remain constant over the entire spatial extent of a given region. Also, it will not remain constant over time. As such, substantial spatial and temporal variations in the concentration of particulates can be expected. It is generally claimed that the impact of particulates on the environment and on man depends on the total amount of exposure over the entire year. Spatial variations can be accounted for by analyzing miniscule units of urbanized regions. This requires extensive calculations based on a diffusion model or a large-scale monitorial program. Since the use of a large-scale monitoring network is infeasible in most situations, the problem can be adequately addressed using diffusion models to predict air quality values over the entire spatial area.

Mitigation of Impact. Particulate pollution impacts can be mitigated by means of four major alternatives:

Reduction in particulate emission from sources.
Reduction or removal of receptors from the polluted areas.
Particulate removal devices such as: cyclones, settling chambers, impactors, scrubbers, electrostatic precipitators, and bag houses.

Uses of protected controlled environment (e.g., oxygen masks, Houston Astrodome).

A combination of the first three alternatives should be considered to provide an optimal strategy for the mitigation of particulate pollution impacts.

Secondary Effects.   Particulate emissions are associated with problems of human health—increased mortality and morbidity in the exposed population. In addition to these direct effects, particulates also cause numerous secondary impacts. Particulates soil clothes and structures, resulting in economic loss. In addition to visibility problems and increased accident risk, aesthetic considerations reduce property values and undermine the general psychic welfare of people. Steel and other metal structures can be corroded as a result of exposure to particulates and humidity. Water quality from storm runoff and vegetation can also be deteriorated by particulate matter present.

Other Comments.   Particulates are present even in the cleanest air at the most remote locations uncontaminated by man. Sources of particulate pollution relate to activities such as construction, industrial operations, operation/maintenance/repair work, and transportation. Automobile emissions are only a minor source of particulate pollution.

Additional References.   "Guidelines for Development of a Quality Assurance Program. Reference Method for the Determination of Suspended Particulates in the Atmosphere—High Volume Method," Environmental Protection Agency, August 1973.

Sulfur Oxides

Definition of the Attribute.   Sulfur oxides are common air pollutants, generated primarily by combustion of fuel. Solid and liquid fossil fuels contain a high degree of sulfur in the form of inorganic sulfides and organic sulfur compounds. Combustion of fossil fuels normally produces about 30 parts sulfur dioxide for 1 part sulfur trioxide.

Sulfur oxides are usually a combination of sulfur dioxide, sulfur trioxide, sulfuric acid, and sulfurous acid. Sulfur dioxide is the most dominant portion of the sulfur oxides concentration; as such, the sulfur oxides attribute is defined in terms of sulfur dioxide parameter.

Sulfur dioxide is a nonflammable, nonexplosive, transparent gas with a pungent, irritating odor. The concentration of this gas in parts per million (ppm) measures the magnitude of sulfur oxides pollution in a given region.

Activities that Affect the Attribute. Many human activities use fossil fuels. Coal and oil-fired furnaces, fossil fueled electric generating plants, and industrial uses of fossil fuels appear to be major generators of sulfur dioxide pollution. In addition, operation of various facilities can cause significant sulfur dioxide pollution. Construction work and transportation also create a minor sulfur dioxide problem from the operation of diesel engines.

Source of Effects. The effects of sulfur dioxide pollution can be higher morbidity, increased mortality, increased incidence of bronchitis, respiratory diseases, emphysema, and general deterioration of health. It can also cause increased corrosion of metals, chronic plant injury, excessive leaf droppings, and reduced productivity of plants and trees. The effect of sulfur dioxide pollution in the presence of particulates can result in synergistic impacts on the environment. Synergistic impacts of sulfur dioxide in the presence of nitrogen dioxide have also been noted. For example, a concentration of 0.04 ppm sulfur dioxide alone does not affect bronchitis or lung cancer patients. However, this concentration of sulfur dioxide combined with 160 $\mu g/m^3$ particulates caused an appreciable increase in mortality of bronchitis and lung cancer patients.[3]

Variables to be Measured. The primary variable that measures the extent of the sulfur oxides problem is expressed by the 24-hour annual arithmetic mean concentration of sulfur dioxide present in the ambient air. The variable is used to predict potential sulfur oxides impact on the environment.

Here, the use of one variable is not entirely adequate. Concentration of particulates, ozone, and nitrogen oxides affects the impacts of sulfur oxides. However, to take advantage of the simplification, only one variable has been used.

How Variables are Measured. The sulfur dioxide concentration is commonly measured by the Pararosaniline method.[2] In principle, sulfur dioxide is absorbed from air in a solution of potassium tetrachloromercurate (TCM). The resulting complex is added to pararosaniline and formaldehyde to form an intensely colored acid solution which is analyzed spectrophotometrically. The spectrophotometric analysis is a colorimetric method in which the concentration of sulfur dioxide absorption is measured by the intensity of the color produced in the resulting acid solution. The method is recommended by the Environmental Protection Agency in the National Primary and Secondary Ambient Air Quality Standards published in the *Federal Register* of April 30, 1971.[2]

Data source. Air quality measurements on sulfur oxide are made by air quality monitoring programs established by state pollution control agencies, the Federal Environmental Protection Agency, and county, regional, multicounty, or city air pollution control agencies. Generally, the data are compiled annually and are published with summaries by the state agency for air quality monitoring.

Skills required. The skills required for measuring sulfur dioxide concentration in air can be developed by special technician-level training imparted at a technical school or as part of an on-the-job training program. Technician-level training in mechanical and chemical engineering is adequate to develop the necessary skills to operate a monitoring and recording system for sulfur dioxide.

Instruments. The instruments required for monitoring sulfur dioxide concentration are:

All-glass midget impinger
Air pump
Air flowmeter
Spectrophotometer.

Evaluation and Interpretation of Data. A review of literature indicates that the minimum sulfur dioxide concentration for vegetation damage is 0.03 ppm. A sulfur dioxide concentration less than 0.03 ppm should be considered a characteristic of a safe environment. As concentration increases, more damage will be done to the vegetation and materials. Visibility of the atmosphere is also impaired. At a concentratoin of 0.2 ppm of sulfur dioxide, increased mortality rates are observed. This situation should reflect a value function of zero. Based on these considerations, a value function was developed for sulfur oxide, as shown in Fig. B-2.

The determination of environmental impact of proposed activities on sulfur dioxide level is measured by the change in sulfur dioxide concentration. When the sulfur dioxide concentration changes to the extent that its rating remains unaltered (i.e., High Quality air remains High Quality, and so on), the impact is considered Insignificant. If the change in sulfur dioxide concentration is such that its rating changes by one step (i.e., from High Quality to Moderate Quality, etc.), the impact is treated as Moderate. Furthermore, when a charge in sulfur dioxide rating occurs through two steps (i.e., from High Quality to Low Quality, and vice versa), the impact is treated as Significant.

The sulfur dioxide value function (Fig. B-2) is used for rating air quality in terms of High, Moderate, and Low Quality based on 24 hour annual arithmetic mean. For a given value of 24 hour annual geometric mean sulfur dioxide concentration on the horizontal axis, the environmental quality rating can be read for the horizontal axis in Fig. B-2.

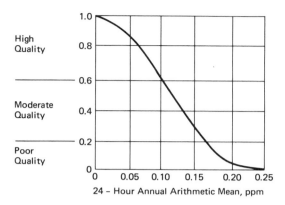

Fig. B-2   Sulfur dioxide value function.

Geographical and Temporal Limitations.   Concentration of sulfur dioxide does not remain constant over the entire spatial extent in a given region. Also, it will not remain constant over time.   As such, substantial spatial and temporal variations in the concentration of sulfur dioxide on the environment and on man depends on the total amount of exposure over the entire year.   Spatial variations can be accounted for by taking miniscule units of urbanized regions for purposes of analysis.   This requires extensive calculations based on a diffusion model or a large-scale monitoring program.   Since the use of a large-scale monitoring network is infeasible in most situations, the problem can be adequately addressed using diffusion models to predict air quality values over the entire spatial region.

Mitigation of Impact.   The impacts can be mitigated by means of four major alternatives or a combination thereof:

Reduction in sulfur dioxide emissions from sources.
Reduction or removal of receptors from the polluted areas.
Gas removal devices using absorption (liquid as a media), adsorption (molecular sieve), and catalytic converters.
Uses of protected, controlled environment, such as oxygen masks, Houston Astrodome, etc.

Secondary Effects.   Secondary effects of sulfur oxides include economic and resource loss through damage to material surfaces and vegetation, water quality deterioration through the natural "cleansing" of the atmosphere through precipitation, and aesthetic and general welfare quality reduction that accompanies the degradation of a vital resource.   Land use patterns and community needs may be affected in localized areas with point source emissions.

Other Comments.   Sulfur dioxide is generally harmful to the health and welfare of a community.   Its impact can be substantially increased by the presence of suspended particulates due to the synergistic relationship of the two pollutants.   Despite this, the value function is based only on the concentration of sulfur dioxide.   This is done to simplify the value function.   However, the impacts have been adjusted for the concentration of particulates that generally accompany given levels of sulfur dioxide in the ambient air.

Additional References.   "Guidelines for Development of a Quality Assurance Program.   Reference Method for Measurement of Sulfur Dioxide," Environmental Protection Agency, August 1973.

Hydrocarbons

Definition of the Attribute.   Hydrocarbon is a general term used for several organic compounds emitted when petroleum fuels are burned.   Automobile exhaust accounts for over half of the complex mixture of hydrocarbons emitted to the atmosphere; the remaining hydrocarbons arise from natural sources like decomposable organic matter on land, swamps, and marshes; hydrocarbon haze from plants and forest vegetation; geothermal areas; coal fields, natural gas, and petroleum fields; and forest fires. Usually, hydrocarbons consist of methane, ethane, propane, and derivatives of aliphatic and aromatic organic compounds.

The hydrocarbons attribute is defined as the total hydrocarbon concentration (THC) present in the ambient air.   Hydrocarbons are organic compounds consisting of carbon and hydrogen; their concentration is measured in parts per million by volume or in micrograms per cubic meter of air. For most U. S. cities, except Los Angeles, the peak hydrocarbon concentration occurs between 6:00 and 9:00 a.m.

Activities that Affect the Attribute.   Many activities emit high levels of hydrocarbons into the environment.   For example, industrial operations involve substantial combustion of fuel, causing hydrocarbon emissions due to inefficient combustion processes.   Gasoline and diesel engines are used for purposes of construction, operation, maintenance, repair, and transportation.   In addition, many industrial activities have petroleum and petrochemical operations that emit high levels of hydrocarbons.   Areas with natural vegetation and forests also generate high levels of hydrocarbon concentration.

Source of Effects.   Hydrocarbons are of concern primarily for their role in the formation of photochemical oxidants and smog. Direct health effects of gaseous hydrocarbons in the ambient air have not been demonstrated. Health effects occur only at high concentrations (about 1000 ppm or more) that interfere with oxygen intake. Hydrocarbons in the atmosphere have been found to cause lacrimation, coughing, sneezing, headaches, nervous weakness, laryngitis, pharyngitis, and bronchitis, even at low concentrations. In addition, hydrocarbons may cause breathing problems and eye irritation. In combination with nitrogen oxides, hydrocarbon impacts can be significantly increased.

Variables to be Measured.   The variable expressing the impact of hydrocarbons is measured by the 3 hour average annual concentration of ambient hydrocarbons, expressed in parts per million. The time concentration is measured from 6:00 to 9:00 a.m., at which time peak hydrocarbon concentration is expected to occur in most U. S. cities except Los Angeles.

Nitrogen oxide variables interact synergistically with the concentration of hydrocarbons. Nitrogen oxides combined with hydrocarbons gernerate oxidants causing smog. The impact of smog is significantly greater than that of hydrocarbons alone. However, for purposes of simplicity, nitrogen oxides are treated as a separate variable.

How Variables are Measured.   There are two different methods of analysis for the total hydrocarbons:

Flame ionization method.[2]
Spectrophotometric method.[4]

The Environmental Protection Agency, in its national primary and secondary ambient air quality standards document, has recommended use of the hydrogen flame ionization method to measure total hydrocarbon concentration. The flame ionization technique uses a measured volume of ambient air delivered semi-continuously (about 4 to 12 times per hour) to a hydrogen flame ionization detector (FID). A sensitive electrometer detects the increase in ion concentration which results from the interaction of hydrogen flame with a sample of air contaminated with organic compounds such as hydrocarbons, aldehydes, and alcohols. The ion concentration response is approximately proportional to the number of organic carbon atoms in the sample. The FID serves as a carbon atom counter.

The measurement can be made by two modes of operation:

A complete chromatographic analysis showing continuous output from the detector.
Programming the system to display selected output from the detector.

The latter is adequate for recording hydrocarbons system concentration values from 6:00 to 9:00 a.m. only.

Data sources. Hydrocarbon data are generally collected by state air quality monitoring programs. Other potential sources include the Federal Environmental Protection Agency, and city or county monitoring agencies.

Skills required. Basic paraprofessional training in mechanical or chemical engineering with special training in operating air pollution samplers is adequate to collect data relating to hydrocarbons. Specialized supervision is needed to insure that the instruments are correctly operated and recorded. This requires either experienced personnel or experienced consultants specialized in air quality monitoring.

Instruments. Instruments used for measuring hydrocarbons are the following:

Commercial Total Hydrocarbon Concentration (THC) analyzer.

Sample introduction system (including pump, flow control valves, automatic switching valves, and flow meter).

In-line filter (a binder-free glass-fiber filter with a porosity of 3 to 5 microns).

Stripper or percolumn (the column should be repacked or replaced every 2 months of continuous use).

Oven (containing analytical column and analytical converter).

The instruments are installed and connected in accordance with the manufacturer's specifications.

Evaluation and Interpretation of Data. The extent of hydrocarbon impact is measured by the degree to which it affects smog intensity. Hydrocarbon criteria are, therefore, keyed to the 6:00 to 9:00 a.m. average annual concentration. At low concentrations, hydrocarbons are relatively harmless and unimportant. The quality of the environment deteriorates rapidly as conditions for smog development approach (i.e., 0.15 ppm to 0.25 ppm). A sharp decrease in environmental quality is noted within this range. Above 0.25 ppm hydrocarbon concentration, the value function gradually levels off to zero, since the marginal impact of increases in hydrocarbons concentration is small. The value function is thus a flat "S" curve. On the basis of these considerations, the hydrocarbons value function shown in Fig. B-3 was developed.

The determination of environmental impact of proposed activities on hydrocarbons level is measured by the change in hydrocarbon concentration. When the hydrocarbon concentration changes to the extent that its rating remains unaltered (e.g., High Quality air remains High Quality), the impact is considered Insignificant. When the change in hydrocarbons concentration is such that its rating changes by one step (e.g., between High Quality and Moderate Quality), the impact is treated as Moderate. When a change in hydrocarbons rating occurs through two steps (e.g., from High Quality to Low Quality, and vice versa), the impact is considered Significant.

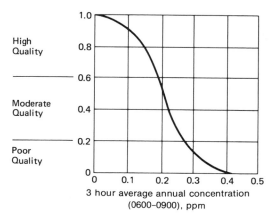

Fig. B-3 Hydrocarbons value function.

The hydrocarbons value function (Fig. B-3) is used for rating air quality in terms of High, Moderate, and Low Quality based on 3 hour average annual concentration. For a given value of 3 hour average annual concentration on the horizontal axis, a point on the curve identifies the environmental quality rating from the vertical axis of Fig. B-3 (e.g., 0.3 ppm indicates a Poor Quality of 0.15).

Geographical and Temporal Limitations. Concentration of hydrocarbons does not remain spatially constant in a given region. Also, it will not remain constant over time. As such, a substantial spatial and temporal variation in the concentration of hydrocarbons can be expected. It is generally claimed that the impact of hydrocarbons on the environment and man depends upon the total amount of exposure during peak periods. Spatial variations can be accounted for by taking small units of urbanized regions for analysis. This requires extensive calculations based on a diffusion model or a large-scale monitoring program. Since the use of a large-scale monitoring network is infeasible in most situations, the problem can be adequately addressed using diffusion models to predict air quality values over the entire spatial area.

Mitigation of Impact. There are four major strategies for the mitigation of impacts of hydrocarbons on the environment. These are:

Control of motor vehicle emissions.
Control of stationary source emission (including evaporation, incineration, absorption, condensation, and material substitution).
Reduction or removal of receptors from polluted areas.
Creation of controlled environment to avoid pollution (including use of oxygen masks).

These strategies can be used in an optimal combination in order to get the best results from an abatement program.

Secondary Effects. Production of hydrocarbons beyond acceptable levels may result in secondary impacts through reduction of property values, shifts in land use patterns, and adverse effects on vegetation. Increased accident occurrence can accompany the direct effects on vision and other human health aspects.

Other Comments. Hydrocarbon concentration is one of the parameters that defines the extent of smog development in an environment. In selecting attributes, the ozone parameter was avoided since the formation of ozone is determined by the interaction of hydrocarbons and nitrogen oxides in the presence of sunlight. The environment receives many different kinds of hydrocarbon emissions; as such, these emissions are an important indicator of environmental impact.

Nitrogen Oxides

Definition of the Attribute. Many nitrogen oxides are found in the urban environment. The most important are nitric oxide (NO) and nitrogen dioxide (NO_2). In addition, nitrous oxide (N_2O) is another oxide of nitrogen present in the atmosphere in appreciable concentration. The term NO_x is often used to represent the composite atmospheric concentration of nitrogen oxides in the environment.

Nitrogen oxides are emitted by exhausts from high temperature combustion sources. They result from the reaction of nitrogen with oxygen; with hydrocarbons they produce photochemical smog. Nitrogen oxide concentrations are measured in parts per million by volume.

Activities that Affect the Attribute. Many human activities generate nitrogen oxides which are emitted to the air. Industrial operations, research/development/testing operations, operation and maintenance of motor vehicles, and stationary combustion sources (like power plants, natural gas burners, diesel-operated construction machineries) are some of the sources of nitrogen oxides. However, a large portion of nitrogen oxides is produced by natural sources, such as bacterial action in forests, swamps, and parks.

Source of Effects. There is very little documented information on the health effects of nitrogen oxides at concentrations normally found in ambient air. The human threshold for sensing the odor of nitrogen dioxide is about 0.12 ppm. Data from human and animal studies indicate that nitrogen oxides have untold effects on human health. Nitrogen dioxide is about four times more toxic than nitric oxide.[5]

In addition, nitrogen oxides can affect vegetation, causing acute (chronic) injury to leaves as well as to productivity of certain plants. Nickel alloys are subject to corrosion in the presence of nitrogen oxides; synthetic fibers fade, and white clothes yellow in the presence of nitrogen oxides.

Variables to be Measured. The variable measuring the extent of nitrogen oxides pollution is the average annual concentration of nitrogen oxides in the ambient air. The nitrogen oxides level is measured in parts per million (ppm).

Other variable factors that might interact with nitrogen oxides are hydrocarbons and particulates. These variables are considered separately in defining air quality impacts, even though they interact synergistically.

How Variables are Measured. Nitrogen dioxide is the only atmospheric nitrogen oxide which can be measured directly with current techniques.* Measurement of nitrogen oxides, therefore, must rely on some type of converter that oxidizes nitric oxide to nitrogen dioxide.

The reference method for the determination of nitrogen dioxide is the Griess-Saltzman technique, modified by the Environmental Protection Agency. It is a 24 hour continuous sampling method. In principle, nitrogen dioxide contaminated air is bubbled through a sodium hydroxide solution, forming a stable solution of sodium nitrite. The nitrite concentration in the sample solution is measured colorimetrically by the reaction of an exposed absorbing agent with phosphoric acid, sulfanilamide, and NEDA solution.

Data sources. Source of data are generally state pollution control departments and county, multicounty, or city air pollution control offices. They can also install monitoring samplers at critical distances from its operation to determine the level of nitrogen oxides generated by its activities.

Skills required. Basic paraprofessional training in mechanical or chemical engineering, with special training in operating air quality sampling devices, is adequate to collect data relating to nitrogen oxides. Specialized supervision is needed to insure that the data are properly collected and analyzed. Specialized supervision should include personnel or experienced consultants trained in the field of air quality monitoring.

Instruments. Nitrogen dioxide is measured with an apparatus consisting of the following instruments:

Absorber tubes
Probe with membrane filter, glass funnel, and trap

---

*The extent of nitrogen oxides pollution is measured by the concentration of nitrogen dioxide expressed in terms of annual arithmetic mean concentration.

Flow control device with a calibrated 27-gauge hypodermic needle and a membrane filter protection
Air pump capable of maintaining a flow of 0.2 liters per minute and a vacuum of 0.7 atmosphere
Calibration equipment.

Evaluation and Interpretation of Data.   Generally, nitrogen oxide concentration below 0.05 ppm (on average annual basis) does not pose a health problem. Exposure above this level can be correlated with a higher incidence of acute respiratory problems. At levels higher than those normally present in ambient air (i.e., about 0.05 ppm), nitrogen dioxide acts as a toxic agent. Based on these considerations, a nitrogen dioxide value function has been developed, as shown in Fig. B-4.

The determination of environmental impact of proposed activities on nitrogen oxides level is measured by the change in nitrogen oxides (NO_x) concentration. When the NO_x concentration changes to the extent that its rating remains unaltered (e.g., High Quality air remains High Quality), the impact is considered Insignificant. When the change in NO_x concentration is such that its rating changes by one step (e.g., between High Quality and Moderate Quality), the impact is treated as Moderate. When a change in NO_x ratings occurs through two steps (e.g., from High Quality to Low Quality, and vice versa), the impact is considered Significant.

The nitrogen oxides value function (Fig. B-4) is used for rating air quality in terms of High, Moderate, and Low Quality, based on average annual concentration. For a given value of average annual concentration on the horizontal axis, a point on the curve identifies the environmental quality rating from the vertical axis of Fig. B-4 (e.g., 0.1 ppm indicates a Poor Quality of 0.1).

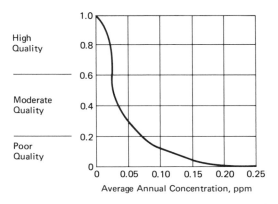

Fig. B-4 Nitrogen oxides value function.

Geographical and Temporal Limitations. Concentration of nitrogen dioxide does not remain constant over the entire spatial extent in a given region. Also, it will not remain constant over time. As such, substantial spatial and temporal variations in the concentration of nitrogen dioxide can be expected. It is generally claimed that the impact of nitrogen dioxide on the environment and man depends on the total amount of exposure over the entire year. Spatial variations can be accounted for by analyzing small units of urbanized regions. This requires extensive calculations based on a diffusion model or a large-scale monitoring program. Since the use of a large-scale monitoring network is infeasible in most situations, the problem can be adequately addressed using diffusion models to predict air quality values over the entire spatial area.

Mitigation of Impact. There are five major strategies for the mitigation of impacts of nitrogen dioxide on the environment. These are:

Control of motor-vehicle emissions
Control of stationary-source emissions (including incineration and evaporation)
Reduction or removal of receptors from polluted areas
Gas removal devices using absorption (liquid as a media), adsorption (molecular sieve) and catalytic converters
Creation of controlled environment to avoid pollution (such as the use of oxygen masks).

These strategies can be used in optimal combination to get the best results from an abatement program.

Secondary Effects. Secondary effects due to nitrogen oxide production include economic loss from damage to vegetation and deterioration of materials. Shifting land use patterns, reduced property values, and increased accident occurrence can accompany the formation of smog and other direct effects.

Other Comments. The standards for nitrogen dioxide are currently being examined by the Environmental Protection Agency, and the value function for nitrogen dioxide could change as a result.

Carbon Monoxide

Definition of the Attribute. Carbon monoxide (CO) is the most widely distributed and most commonly occurring air pollutant. The majority of atmospheric CO is produced by the incomplete combustion of carbonaceous materials used for fuels for vehicles, space heating, industrial processing, and the burning of refuse.

Activities that Affect the Attribute.   All activities that involve the combustion of organic materials are sources of CO. In addition, industrial operations also contribute to the CO burden of the air. CO is also formed by explosions, firing of weapons, and can occur naturally.

Source of Effects.   Adverse health effects on humans have been observed for exposures of 8 hours or more at CO concentrations of 12 to 17 milligrams per cubic meter (10 to 15 ppm). Adverse health effects consist of impaired time interval discrimination, physiologic stress on heart patients, etc.

There is, however, no evidence that CO has adverse effects on vegetation or materials.

Variables to be Measured.   The concentration of CO is measured in micrograms per cubic meter. The variable measuring the extent of carbon monoxide pollution is the maximum 8 hour concentration of CO that has been recorded over a period of 1 year.

How Variables are Measured.   The reference method for the continuous measurement of carbon monoxide is nondispersive infrared spectrometry. The measurement technique is based on the absorption of infrared radiation by carbon monoxide. By comparing absorption of infrared radiation passing through a reference cell and a test cell electronically, the concentration of CO in the test cell can be measured.

Instruments are available that measure in the range of 0 to 58 mg/m^3. The sensitivity is 1 percent of full scale response per 0.6 mg CO/m^3 (0.5 ppm).

Data sources.   The sources of data are generally the State Pollution Control Department, the County Air Pollution Control Office, or the City Air Pollution Control Office. Monitoring equipment can be installed at critical locations near specific operations to determine the level of carbon monoxide generated.

Skills required.   A basic paraprofessional training in mechanical or chemical engineering with special training in operating the air quality instruments is adequate to collect data relating to carbon monoxide. Specialized supervision will be needed to insure that the data are properly collected and analyzed. Specialized supervision should include trained and experienced personnel or experienced consultants in the field of air quality monitoring.

Instruments.   Instruments recommended for measuring carbon monoxide are:

Commercial nondispersive infrared spectrometer
Sample Introduction System (including pump, flow control valve, and flowmeter)
In-line filter (a filter with a porosity of 2 to 10 microns should be used to trap large particles)
Moisture controller (refrigeration units or drying tubes).

The instruments are installed and connected in accordance with the manufacturer's specifications.

Evaluation and Interpretation of Data. Generally, carbon monoxide does not pose a health problem to the general public. Continuous exposure to CO concentrations of 10-15 ppm, however, can cause impaired time interval discrimination. CO levels of 30 ppm have caused physiologic stress in patients with heart disease, while concentrations of 8 to 14 ppm have been correlated with increased fatality rates in hospitalized myocardial infarction patients.

Geographical and Temporal Limitations. The concentration of carbon monoxide does not remain constant over the entire spatial extent in a given region. Also, it will not remain constant over time. As such, substantial spatial and temporal variations in the concentration carbon monoxide can be expected. It is generally claimed that the impact of carbon monoxide on the environment and man depends on the total amount of exposure over the entire year. The spatial variations can be accounted for by taking small units of urbanized regions for purposes of analysis. This would require extensive calculations based on a diffusion model or a large-scale monitoring program. Since the use of a large-scale monitoring network is infeasible in most situations, the problem can be adequately addressed using diffusion models to predict air quality values over the entire spatial area.

Mitigation of Impact. There are three major strategies for the mitigation of impact of carbon monoxide on the environment. These are:

Control of motor vehicle emissions
Control of stationary source emission
Reduction or removal of receptors from polluted areas.

Secondary Effects. Presently identifiable specific secondary impacts due to increased carbon monoxide emissions are those related to human health effects—economic loss, increased accident rate, etc. Long-term secondary effects on the ecosystem due to increased carbon monoxide levels are not yet understood.

Additional References. "Guidelines for Development of a Quality assurance Program. Reference Method for Continuous Measurement of Carbon Monoxide in the Atmosphere," Environmental Protection Agency, August 1973.

Photochemical Oxidants

Definition of the Attribute. Products of atmospheric reactions between hydrocarbons and nitrogen oxides which are initiated by sunlight are called photochemical oxidants.   The product of these reactions which is most commonly found and measured in the atmosphere is ozone.   Other oxidants of interest include peroxyacetyl nitrate (PAN) and acrolein.   Atmospheric measurement techniques measure the net oxidizing properties of atmospheric pollutants and report these photochemical oxidant concentrations as equivalent ozone concentration.   Photochemical oxidants can be found anywhere where hydrocarbons and nitrogen oxides interact in the presence of sunlight.

Activities that Affect the Attribute.   All activities that generate oxides of nitrogen and hydrocarbons simultaneously contribute to the generation of photochemical oxidants.   Industrial activities and the operation and maintenance of motor vehicles and stationary combustion sources are major sources of nitrogen oxides and hydrocarbons.   In addition, many other activities have petroleum and petrochemical operations that emit high levels of hydrocarbons.

Source of Effects.   The data from animal and human studies are sparse and inadequate for determining the toxicological potential of photochemical oxidants.   Injury to vegetation is one of the earliest manifestations of photochemical air pollution.   The oxidants can cause both acute and chronic injury to leaves.   Leaf injury has occurred in certain sensitive species after a 4 hour exposure to 100 micrograms per cubic meter (0.05 ppm) total oxidants.   Photochemical oxidants are known to attack certain materials.   Polymers and rubber are important materials that are sensitive to photochemical oxidants.

Variables to be Measured.   The variable measuring the extent of photochemical oxidants is the maximum 3 hour concentration (6:00 to 9:00 a.m.) not to be exceeded more than once a year.   The photochemical oxidant level is reported in micrograms per cubic meter.

How Variables are Measured.   Since ozone is the major constituent contributing to photochemical oxidants, it is used as the reference substance in reporting levels of photochemical oxidants.
    Ambient air and ethylene are delivered simultaneously to a mixing zone where the ozone in the air reacts with the ethylene to emit light which is detected by a photomultiplier cell.   The resulting photocurrent is amplified and displayed on a recorder.   The range of most instruments is from 0.005 ppm to greater than 1 ppm of ozone.   The sensitivity is 0.005 ppm of ozone.
    Data sources.   The sources of data are generally the State Pollution Control Department, the County Air Pollution Control Office, or the City Air Pollution Control Office.   They can also install monitoring equipment at critical locations near its operations to determine the level of photochemical oxidants generated by its activities.

Skills required. A basic paraprofessional training in mechanical or chemical engineering with special training in operating the air quality instruments is adequate to collect data relating to photochemical oxidants. Specialized supervision will be needed to insure that the data are properly collected and analyzed. Specialized supervision should include trained and experienced personnel or experienced consultants in the field of air quality monitoring.

Instruments. Instruments for carrying out the photochemical oxidant measurement include:

A detector cell

An air flowmeter capable of controlling air flows between 0-1.5 liter per minute

An ethylene flowmeter capable of controlling ethylene flows between 0-50 milliliter per minute

An air inlet filter capable of removing all particles greater than 5 microns in diameter

A photomultiplier tube

A high-voltage power supply (2000 volts)

A direct-current amplifier and a recorder.

Evaluation and Interpretation of Data. Photochemical oxidants are keyed to the 6:00 to 9:00 a.m. concentration values. At low concentrations, photochemical oxidants do not pose a problem. The quality of the environment, however, rapidly deteriorates as conditions for smog development approach, i.e., hydrocarbon concentrations of 0.15 to 0.25 ppm. The values of the oxidant levels during the early morning determine the intensity of the oxidants to be expected later in the day. After sunset, the oxidant concentrations are reduced to low levels.

Geographical and Temporal Limitations. The concentration of photochemical oxidants does not remain constant over the entire spatial extent in a given region. Also, it will not remain constant over time. As such, a substantial spatial and temporal variation in the concentration of photochemical oxidants can be expected. It is generally claimed that the impact of photochemical oxidants on the environment and man depends on the total amount of exposure during the peak periods. The spatial variations can be accounted for by taking small units of urbanized regions for purposes of analysis. This would require extensive calculations based on a diffusion model or a large-scale monitoring program. Since the use of a large-scale monitoring network is infeasible in most situations, the problem can be adequately addressed using diffusion models to predict the air quality values over the entire spatial area.

Mitigation of Impact. All strategies for mitigating hydrocarbons and oxides of nitrogen are applicable to photochemical oxidants.

Secondary Effects.   Sensitivity of plants to photochemical oxidants results in economic loss, as well as other secondary impacts on ecological balance. Other economic loss occurs with material deterioration and reduced property values.

Additional References.   "Guidelines for Development of a Quality Assurance Program. Reference Method for the Measurement of Photochemical Oxidant," Environment Protection Agency, August 1973.

Hazardous Toxicants

Definition of the Attribute.   Many kinds of hazardous air pollutants may be released to the environment. Some of these toxic elements or compounds are arsenic, asbestos, barium, beryllium, boron, cadmium, chromium, copper, lead, molybdenum, nickel, palladium, titanium, tungsten, vanadium, zinc, zirconium, radioactive wastes, mercury, and phenols. These toxic substances at certain concentrations may cause serious damage to the health and welfare of an exposed community.

Hazardous toxicants are substances like asbestos, beryllium, mercury, other harmful elements, and their compounds. Man's exposure to these toxicants can cause serious health hazards and diseases. These health impairments can result in increased mortality, morbidity, susceptibility to diseases, and loss of productivity.

Activities that Affect the Attribute.   Hazardous toxicants may be generated by human activities such as construction, operation/maintenance/repair of existing systems, industrial operations, research/development/testing operations, and demolition of structures. For example, the surfacing of roadways with asbestos tailings can cause serious asbestos hazards.

The manufacture of clocks, cord, wicks, tubing, tape, twine, rope, thread, cement products, fireproofing and insulating materials, friction products, paper, mill-board, felt, floor tile, paints, coatings, caulks, adhesives, sealants, and plastics may produce visible emissions of asbestos. Also, construction emissions produce substantial amounts of asbestos dust.

Source of Effects.   Hazardous toxicants can create serious health hazards and diseases of a chronic nature. For instance, exposure to asbestos dust at high concentrations and for longer durations can cause asbestos and bronchial cancer.[8] In addition, asbestos is a cause of mesothliomas; tumor; and membrane, intestine, and abdomen cancers. Most asbestos diseases have a latency period of 30 years.

To date, research has failed to establish an emission limit or concentration range above which asbestos dust can be harmful to human health. The

Environmental Protection Agency, however, recommends that no visible emissions be permitted from asbestos-generating activities.

Beryllium is another hazardous air pollutant which can seriously affect human health. These effects are acute and chronic lethal inhalation, skin and conjuctival effects, cancer induction, and other beryllium diseases. The lowest beryllium concentration producing a beryllium disease was found to be greater than 0.01 microgram per cubic meter.[9] At a concentration of 0.10 microgram per cubic meter or above, a large majority of exposed persons will develop beryllium diseases.

Variables to be Measured.   The variable measuring the extent of impact of a specific hazardous toxicant is given by the maximum concentration of a given toxicant averaged over a 30 day period. For example, beryllium concentrations are required not to exceed 0.01 microgram per cubic meter over a 30 day period. In the same vein, mercury concentrations should not exceed 1.0 microgram per cubic meter over a 30 day averaging period.

How Variables are Measured.   There are many different methods of measuring various hazardous toxicants. The standard methods for sampling hazardous toxicants in the ambient air are presented in the Standard Methods for Sampling Stack Particulate Matter.[10] Also, specific measurement techniques for beryllium and mercury are given in the *Federal Register*, April 6, 1973.

Data sources.   Only a few city, county, regional, and state agencies monitor hazardous toxicants and emissions. Monitoring of selected hazardous toxicants is occasionally done by the Environmental Protection Agency in cooperation with state or local agencies for selected periods at critical locations. Such monitoring is done only when a special hazardous toxicant is identified in a given region. Data on toxicant monitoring are available from state and local air pollution control agencies when collected.

Skills required.   Skills required for the measurement of various hazardous toxicants measuring techniques are not well defined in the literature and require specialized supervision for use. Specialized consulting services are needed to implement these measurement techniques.

Instruments.   Complex sampling trains have to be designed on a case-by-case basis for each hazardous toxicant in the environment. The full range of instrumentation necessary for measurement of each hazardous toxicant is described in some of the standard documents mentioned above.

Evaluation and Interpretation of Data.   There are no well-defined value functions available for the hazardous toxicants identified in the environment. Generally, for each hazardous toxicant, it is possible to establish the upper and lower concentration limit of acceptability for the environment. The upper limit of acceptability is called the permissible level, the excess of which is considered highly undesirable and damaging to human health. On the other hand, the lower concentration limit of acceptability is called the desirable level, below which concentration of the quality of air can be considered ideal; that is, the value function equals 1.

Permissible and desirable limits have not generally been established for each of the known hazardous toxicants. The federal government has established standards for three major hazardous toxicants: asbestos, beryllium, and mercury. The levels for each of these pollutants are as shown in Table B-1.

The environmental impact of proposed activities on hazardous toxicant level is measured by the change in the hazardous toxicant concentration (HTC). When the HTC changes to the extent that its rating remains unaltered (e.g., High Quality air remains High Quality), the impact is considered Insignificant. When a change in HTC is such that the rating changes by one step (e.g., between High Quality and Moderate Quality), the impact is treated as Moderate. When a change in HTC rating occurs through two steps (e.g., from High Quality to Poor Quality or vice versa), the impact is considered Significant.

The hazardous toxicant value function, shown in Table B-1, is used to determine what quality rating should be given to a specific value of HTC. The following quality ratings are given to the HTC values:

Quality Rating	*HTC Values*
High Quality	Desirable level to less than Permissible level
Moderate Quality	Permissible level to less than Dangerous level
Poor Quality	Dangerous level

Geographical and Temporal Limitations. Concentration of hazardous toxicants does not remain constant over the entire spatial extent in a given region. Also, it will not remain constant over time. As such, substantial spatial and temporal variations in the concentration of hazardous toxicants

TABLE B-1 Hazardous Toxicant Value Functions

HTC Values (Value Function)	Hazardous Toxicant		
	Asbestos	Beryllium mg/m^3	Mercury mg/m^3
Permissible level (0.5)	No visible emissions	0.10	.0.1
Desirable level (1.0)	0.0	0.01	0.0
Dangerous level (0.0)	Visible	>0.10	>1.0

can be expected. It is generally claimed that the impact of hazardous toxicants on the environment and man depends on the total amount of exposure over the entire day. Spatial variations can be accounted for by analyzing small units of urbanized regions. This requires extensive calculations based on a diffusion model or a large-scale monitoring program. Since the use of a large-scale monitoring program is infeasible in most situations, the problem can be adequately addressed using diffusion models to predict air quality values over the entire spatial area.

Mitigation of Impact.   There are five major strategies for the mitigation of impacts resulting from hazardous toxicants:

Removal of hazardous emissions
Use of materials that do not generate hazardous toxicants
Use of processes that do not generate hazardous toxicants
Avoiding or reducing activities that generate hazardous toxicants
Removing people from contaminated areas.

Secondary Effects.   Secondary effects of hazardous toxicants include economic loss which accompanies lowered health standards and decreased productivity. Deterioration of water quality may result as these toxicants are cleansed from the air by natural processes. Effects on plant and animal life (aquatic and terrestrial) would vary with the toxicants and the levels present.

Other Comments.   Hazardous toxicants are powerful damaging agents for a community. Any industry can ill afford to be negligent about such emissions. Any attempt on the part of industry to compel communites to endure dangerous levels of toxicants resulting from its activities should be strongly discouraged. The use of this parameter will help to identify potential hazardous toxicant problems resulting from various operations.

Odors

Definition of the Attribute.   Industrial malodors are generally considered harmless, even though they frequently cause loss of personal and community pride, loss of social and economic status, discomfort, nausea, loss of appetite, and insomnia. It is true that odor effects on human health and welfare have been recognized only recently, and it seems that very little attention has been given in the literature to this air contaminant.

Malodors are generally caused by organic and sulfur compounds. The resulting odor characterisitics are described by commonly accepted odor descriptors. Some common odor descriptors and their odor contaminants are indicated in Table B-2. For each odor contaminant, a concentration can be defined for which there can be no perception of the odor by a panel of individuals. The concentration is generally known as olfactory threshold or odor threshold. The odor thresholds of a few selected gaseous sulfur compounds in the air are shown in Table B-3.

TABLE B-2   Selected Malodors and Contaminants

Chemical Compound or Type Material	Commonly Accepted Description of Odor Types
Acetaldehyde	Fruity
Acetic acid	Vinegar
Acetone	Nail polish remover
Acetylene	Ethereal, garlic
African fiber	Musty, sour
Banana oil	Nail polish remover
Burnt protein	Burnt toast, scorched grain
Cannery waste	Rotten egg
Carbon disulfide	Rotten egg
Carbon tetrachloride	Cleaning fluid
Cresol	Creosote
Decayed fish	Rendering
Dimethyl sulfide	Rotten vegetables
Enamel coatings	Fatty, linseed oil
Fatty acids	Grease, lard
Fermentation	Yeast or stale beer
Foam rubber curing	Sour sulfides
Gas house	Gas odors
Hydrogen sulfide	Rotten egg
Indole	Rest room
Iodoform	Iodine
Medicinal	Iodoform
Methyl ethyl ketone	Nail polish remover
Mercaptans (methyl)	Rotten cabbage
Oils: castor, coconut, soya, linseed	Rancid grease
Phenolic	Carbolic acid
Phenolic resins	Carbolic acid
Pig pen	Waste lagoons
Pyridine	Acrid, goaty
Septic sewage	Rotten egg
Skatole	Rest room
Sludge driving	Burnt grain
Sulfur dioxide	Irritating, strong suffocating

Sources: Weisburd[6] and Post[7].

TABLE B-3 Odor Thresholds of Selected Gaseous Sulfur Compounds in Air, ppm (by volume)

Sulfur dioxide	1.0–5.0[a]		
Hydrogen sulfide	0.0047[a]	0.0085[b]	0.0009[d]
Methyl mercaptan	0.0021[a]	0.040[c]	0.0006[d]
Dimethyl sulfide	0.0001[a]	0.0036[c]	0.0003[d]

[a]Leonardis, G., D. Kendall, and N. Barnard, "Odor Threshold Determinators of 53 Odorant Chemicals," *J.A.P.C.A.* **2**: 91–5 (1969).
[b]Lederlof, R., M. L. Edfor, L. Friberg, and T. Lindvall, *Nordisk Hygenish Tidskrfit* pp. 46, 51, 1965.
[c]Young, F. A., D. R. Adama, and D. Sullivan, "The Relationship Between Environmental-Demographic Variables and Olfactory Detection and Objectionability Thresholds," to be published in *Perception and Psychophysics*.
[d]Sheehy, J. P., W. Achinger, R. A. Simon, AP44 "Handbook of Air Pollution," U.S. DHEW, PHS, Bureau of State Services, Division of Air Pollution, Cincinnati, Ohio, 1968.
SOURCE: Weisburd[6].

The odor intensity is a measure of the stimulus resulting from the olfactory sensation of a given concentration of odorant. According to the Weber-Fechner law, odor intensity increases only logarithmically with the increase in concentration of the odor.

Activities that Affect the Attribute. In general, industrial operations, research/development/testing operations, and operation/maintenance activities are potentially capable of emitting odor contaminants to the air.

Specific examples include metallurgical, chemical, petroleum, and food processing operations, feedlots, and burning activities.

Source of Effects. Malodors can affect both health and welfare of a community. These effects result from the loss of personal and community pride, reducing property values, tarnishing silver and paints, corroding steel, reducing appetite, producing nausea and vomiting, causing headache, and disturbing sleep, breathing, and olfactory sensation. These result in significant impacts, causing major public concern.

Variables to be Measured. There are two major variables that measure the extent of odor problems. First, the average annual concentration of selected odor contaminants in parts per million (ppm) by volume is a useful measure of the extent of odor pollution at a given receptor point in a community. Second, the odor intensity, determined by an "odor jury" consisting of a panel of eight persons, is another measure of odor problems. The odor intensity scale has the following levels:

Levels	Descriptors
0	No odor
1	Odor threshold (or very slight odor)
2	Slight odor
3	Moderate odor
4	Strong odor

The concentration and intensity variables are used interchangeably for odor measurements.

How Variables are Measured. Two distinct methods for measuring malodors are:

Scentometer method
Odor judgment panel.

A scentometer can be used to measure ambient odor intensities when traveling through dusty areas. Strong, constant odors are measured by a scentometer over a square mile of area. It is useful routine surveillance device that can identify threshold levels, possible odor-problem areas, patterns of peak odor intensity, etc., over a given region.

On the other hand, an odor judgment panel can be used to verify the source of an unidentified odor, odor intensity, and damage potential of a given odor.

Data sources. The Federal Government has not yet established standards for potential odorants. No systematic monitoring and data collection are done with regard to odorants or odor contaminants by state and local agencies. Only in isolated cases will it be possible to find data on odor contaminants for selected periods and monitoring stations operated by state or local agencies.

Skills required. The use of a scentometer requires at least a technician-level training and about a year's experience in using the equipment. The odor-panel approach does not require any specific qualifications or formal training. It requires careful selection of juries based on olfactory sensation, and continuous training of the jurors to develop proper perception of different types of odors.

Instruments. The scentometer is the only equipment that is required in the first method of measuring odor problems. The second method (i.e., odor-panel approach) does not require any equipment whatsoever.

Evaluation and Interpretation of Data. An environment with no odor at all is considered to be an ideal environment, with an environmental quality value of 1.0. Odor threshold concentration represents a tolerable level of odor contamination in the air; as such, it has an environmental quality value of 0.6. The value function falls rapidly with the occurrence of slight odor, and to 0 with a strong odor. Based on above considerations, the value function for various odorants is presented in Fig. B-5. For practical purposes, odor threshold of any odorant is the odorant concentration that can be detected only by 5 to 10 percent of the panelists. The slight odor is detected by about 20 to 25 percent of the panelists. The moderate odor is detected by about only 40 percent of the panelists, and the strong odor is detected by about 100 percent of the panelists.

Geographical and Temporal Limitations. Concentration of malodors does not remain constant over the entire spatial extent in a given region. Also, it will not remain constant over time. As such, substantial spatial and temporal variations in the concentration can be expected. Spatial variations can be accounted for by analyzing small units of urbanized regions. This requires extensive calculations based on a diffusion model or a large-scale monitoring program. Since the use of a large-scale monitoring network is infeasible in most situations, the problem can be adequately addressed using diffusion models to predict air quality values over the entire spatial area.

The environmental impact of proposed activities on odor level is measured by the change in odor intensity. When the odor intensity changes to the extent that its rating remains unaltered (e.g., High Quality air remains High Quality), the impact is considered Insignificant. When the change in odor intensity is such that its rating changes by one step (e.g., between High Quality and Moderate Quality), the impact is treated as Moderate. When a change in odor rating occurs through two steps (e.g., from High to Low Quality, or vice versa), the impact is considered Significant.

The odor value function (Fig. B-5) is used for rating air quality in terms of High, Moderate, and Low Quality, based on measured odor intensity. For a given value of odor intensity on the horizontal axis, a point on the curve can be found which identifies the environmental quality rating from the vertical axis of Fig. B-5 (e.g., odor intensity of slight odor indicates a Moderate Quality of greater than 0.2).

Mitigation of Impact. The many different methods of abating potential impacts of odorous contaminants include:

Dilution of odorant (dilution can change the nature as well as strength of an odor)
Odor counteraction or neutralization (certain pairs of odors in appropriate concentrations may neutralize each other)
Odor masking or blanketing (certain weaker malodors may be suppressed by a considerably stronger good odor)
Reduction in odor emissions

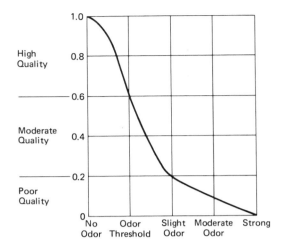

Fig. B-5 Odor value function.

Removal of receptors from polluted areas and/or from downwind odor path

Fatiqued olfactory odor perception (certain levels of odor can be tolerated as a result of perception fatigue due to long-term exposure to the odor).

Planning can help establish optimal combinations of these mitigation alternatives to ensure that the best solution is made available to a community.

Secondary Effects.   Additional effects of malodors include the lowering of socioeconomic status, damaging community reputation, discouraging capital investment in a community, and discouraging tourism.   Effects on the ecosystem and animal populations are not well understood.

REFERENCES

1. Jain, R. K. *et. al.*, "Environmental Assessment Impact Study for Army Military Programs," U. S. Army Construction Engineering Research Laboratory, Technical Report CERL-IR-D-13 December 1973.

2. Environmental Protection Agency, "National Primary and Secondary Ambient Air Quality Standards," Federal Register **36,** *No. 228* (November 25, 1971).

3. *Air Quality Criteria for Sulfur Oxide*, U. S. Department of Health, Education and Welfare, Washington, D.C., March 1970.

4. *Air Quality Criteria for Hydrocarbons*, U. S. Department of Health, Education and Welfare, Washington, D.C., March 1970.

5. *Air Quality Criteria for Nitrogen Oxides*, U. S. Department of Health, Education and Welfare, Washington, D.C., March 1970.

6. Weisburd, N. I., *Field Operations and Enforcement Normal for Air Pollution Control* 3, Environmental Protection Agency, August 1972.

7. Post, N. K., "Odor Control of Wastes," *Industrial Waste Disposal*, R. D. Ross (ed.), Van Nostrand Reinhold, New York, 1968.

8. Environmental Protection Agency, "National Emission Standards for Hazardous Air Pollutants," *Federal Register* 38, *No. 66* (April 6, 1973).

9. Eisenbud, N. *et. al.*, "Nonoccupational Berylliosis," *Ontological Hygiene and Toxicology* 31 (1949).

10. "Standard Method for Sampling Stack Particulate Matter," *1971 Book of ASTM Standards,* Part 23, Philadelphia, Pennsylvania (ASTM, Designation 2928-71), 1971.

11. Environmental Protection Agency, "Standards of Performance for New Stationary Sources," Federal Register 36, *No. 247* (December 23, 1971).

12. Environmental Protection Agency, "Standards of Performance for New Stationary Sources," Federal Register 38, *No. 111* (June 11, 1973).

WATER

Water means different things to different people. A particular definition depends, in large measure, on the personal uses to which water is put by the definer. Water may be considered an absolute necessity to sustain life and a necessary resource for all economic activity by some, and yet a refuge for biological pests and nuisances by others.

Pollution of water is impairment of water quality by man's activity causing an actual hazard to public health or impairment of beneficial use of water.

The water environment is an intricate system of living and nonliving elements. Physical, chemical, and biological factors influencing water quality are so interrelated that a change in any water quality parameter triggers other changes in a complex network of interrelated variables. Often it is difficult to categorize the nature of these interrelationships that may result from man's activity and influence on the entire water system.

To simplify analysis in the area of water, attributes of similar nature have been grouped together. This grouping was done with the following objectives. The list of selected attributes should be:

(1)   As compact as possible
(2)   Equally applicable to surface and groundwater quality

(3)   Representative of comprehensive water quality indicators
(4)   Measurable in the field
(5)   Relevant to the spectrum of major activities
(6)   Capable of being measured on a project scale.

Self-Purification of Natural Waters.   All natural waters have the capability to assimilate certain amounts of waste without apparent effect upon the environment.   The process by which self-purification is achieved is different for surface water and groundwater systems.   Both types of water systems are briefly described below.

Surface Water System.   Some minor degradation of surface water quality may be overcome by the natural capacity of water bodies for withstanding certain insults.   Such natural capacity is a result of dilution, sedimentation, flocculation, volatilization, biodegradation, aeration, aging, and uptake by organisms.   The effects of relatively small amounts of waste are mitigated and the water system recovers itself.   If the waste load is excessive, even for a short period, the effects may be devastating.   The process of self-purification in surface waters is a complex phenomenon.   Readers are referred to several excellent sources of information.[4,5,6]

Groundwater System.   Pollution of groundwater systems is relatively difficult.   Contaminants have to travel through a soil column before any pollution is caused.   Many soils have the capacity to mitigate manifold types of wastes.   The processes by which waste is purified in the soil column are aerobic and anaerobic decomposition, filtration, ion exchange, adsorption, absorption, etc.   The process of dilution also reduces the concentration of contaminants.

Many contaminants are frequently removed during movement of water through the soil (unless the groundwater is directly contaminated by fissure cracks, leaks, pipes, or holes).   Examples of such contaminants are micro-organisms, organic matter, and turbidity.   Only dissolved solids and gases are of significant importance in groundwater pollution.   These contaminants, as discussed later, cause taste, odor, and physiological effects.

When groundwater becomes contaminated, water purification is a difficult problem.   Due to the relatively low flow rates of groundwater systems, pollutants are not readily diluted, and thus, tend to remain localized problems for a period of time.   There is also a considerable lag in time before that pollution becomes noticeable in a groundwater system.   As a result, today's activity may show impact only after several years.

Description of Selected Water Quality Attributes.   Fourteen attributes define potential effects on water quality from the basic activities associated with Army programs. These attributes, in three major categories, physical, chemical, or biological, are outlined below.

I   Physical
    A.   Aquifer safe yield
    B.   Flow variations
    C.   Oil
    D.   Radioactivity
    E.   Suspended solids
    F.   Thermal pollution
II   Chemical
    A.   Acid and alkali
    B.   Biochemical oxygen demand (BOD)
    C.   Dissolved oxygen (DO)
    D.   Dissolved solids
    E.   Nutrients
    F.   Toxic compounds
III   Biological
    A.   Aquatic life
    B.   Fecal coliforms

Table B-4 is a summary table indicating the 14 water quality attributes, conditions contributing to each, and a useful scale of impacts.

Aquifer Safe Yield

Definition of the Attribute.   Aquifer* safe yield describes the general availability of the total groundwater system to supply water for human uses without the ultimate depletion of the aquifer. Aquifer safe yield includes all physical attributes of aquifer, which are porosity, permeability, transmissibility (which is permeability times thickness of the aquifer), and storage coefficient.

Activities that Affect the Attribute.   Many human activities affect the aquifer yield. The aquifer safe yield (available water resource) may decrease due to overpumping or by restricting this movement of water into or through the aquifer. During overpumping as a result of turbulences in the well bore, fine-grained material moving near the well causes a decrease in water movement toward the well. Land use patterns may significantly reduce the water percolation into the ground. Also, improper waste injection may cause clogging of the formation due to suspended solids or bacterial action.

---

*An aquifer may be defined as an earthy material capable of yielding water to a well in usable quantities.

TABLE B-4. Selected Attribute and Environmental Impact Categories

Selected Attributes	Observed Condition	Environmental Impact Category[1]				
		1 (Most Desirable)	2	3	4	5 (Least Desirable)
Physical Aquifer safe yield[2]	Changes occurring in physical attributes of aquifer (porosity, permeability, transmissibility, storage coefficient, etc.)	No change	No change	Slight change	Significant change	Extensive change
Flow variation[3]	Flow variation attributed to activities; (Q_{max}/Q_{min})	None	None	Slight	Significant	Extensive
Oil[4]	Visible silvery sheen on surface, oily taste and odor to water and/or to fish and edible invertebrates, coating of banks and bottom or tainting of attached associated biota	None	None	Slight	Significant	Extensive

		Equal to or less	Equal to or less	Exceed limit	Exceed limit	Exceed limit
Radioactivity[4]	Measured radiation limit 10^{-7} micro curie/ml					
Suspended solids[3]	1. Sample observed in a glass bottle	Clear	Clear	Fairly clear	Slightly turbid	Turbid
	2. Turbidity in Jackson turbidity units	3 or less	10	40	60	140
	3. Suspended solids mg/l	4 or less	10	15	20	35
Thermal discharge[3]	Magnitude of departure from natural condition C	0	2	4	6	10
Chemical						
Acid and alkali[4]	Departure from natural condition, pH units	0	1	2	3	4
BOD[4]	mg/l	1	2	3	5	10
DO[3]	% saturation	100	85	75	60	Low
Dissolved solids[4]	mg/l	500 or less	1000	2000	5000	High
Nutrients[3]	Total Phosphorus mg/l	0.02 or less	0.05	0.10	0.20	Large
Toxic compounds[4]	Concentration mg/l	Not detected	Traces	Small	Large	Large
Biological						
Fecal coliforms[4]	Number per 100 ml	50 or below	5000	20,000	250,000	Large

TABLE B-4. *(Continued)*

Selected Attributes	Observed Condition	Environmental Impact Category[1]				
		1 (Most Desirable)	2	3	4	5 (Least Desirable)
Aquatic life[3]	Green Algae	Scarce	Moderate quantities in shallows	Plentiful in shallows	Abundant	Abundant
	Grey Algae	Scarce	Scarce	Scarce	Present	Plentiful
	Delicate fish; trout, grayling	May be plentiful	Plentiful	Probably absent	Scarce	Absent
	Coarse fish; chub, dace, carp, roach	May be present	Plentiful	Plentiful	Scarce	Absent
	Mayfly naiad, stonefly nymph	May be plentiful	Plentiful	Scarce	Absent	Absent
	Blood worm, sludge worm, midge larvae, rat-tailed maggot, sewage fly larvae and pupa	May be absent	Scarce	May be present	Plentiful	Abundant

Notes: (1) Environmental Impact Category; Category 1 indicates most desirable condition; Category 5 indicates extensive adverse condition. Because all attributes are related to environmental quality between 0 and 1 it is possible to compare different attributes and five categories on a common base. Each category is equivalent to approximately 20% of overall environmental quality. In the physical sense, water quality for five categories will be very clean, clean, fairly clean, doubtful and bad. Environmental impact may be adverse or favorable. Adverse impact will deteriorate the environmental quality while favorable impact will improve the quality. Proper signs and weights must be used to achieve overall effects.

(2) Applies to ground water systems only.
(3) Applies to surface water systems only.
(4) Applies to both the ground water and surface water.

Leaching of landfills may also clog the pores. All these factors decrease transmissibility of an aquifer and result in decreased aquifer safe yield. In regions dependent upon groundwater for water supplies, a decrease in safe yield could be highly undesirable. Lowering of the water table may cause public controversy, even in regions almost wholly dependent upon surface waters as a water supply. In coastal regions, uncontrolled water pumping from the ground may reverse the normal seaward gradient of the water table and permit saltwater to move inland and contaminate the aquifer.

Many activities may increase water availability due to increased water entering the system, which may result in raising of the water table accompanied by increased aquifer safe yield. Examples of such activities are water impoundment and reservoir construction and changes in topography to increase percolation. High water table is often accompanied by water-logging problems in soils and water problems during excavation.

Source of Effects. As discussed above, many activities may upset the aquifer yield by directly or indirectly altering physical factors such as permeability, porosity, and ground surface conditions. The effects may be damaging and reduce potential groundwater resources.

Variables to be Measured; How Variables are Measured. Maximum safe yield is measured in thousands of acre-feet of water withdrawn in a unit of time (usually in a year); the method of measurement is based upon several techniques which all utilize extensive pumping tests.

Evaluation and Interpretation of Data. Knowledge concerning the relationship between degree of change in aquifer safe yield and the environmental impact is extremely limited. It would not be possible at this time to make any quantitative judgement. However, since the reasonable environmental goal is to minimize the impact, a qualitative judgment can be made relating to deviation from the natural condition. Table B-4 summarizes five degrees of environmental impacts based upon the qualitative judgments.

Geographical and Temporal Limitations. Impacts related to aquifer safe yield are most likely to occur in areas (1) with high dependency on groundwater for supply; (2) with a high water table; or (3) with significant seasonal precipitation and subsequent infiltration. Local U. S. Geological Survey offices are excellent data sources for groundwater information.

Mitigation of Impact. All activities likely to change the physical nature of the aquifer, to affect land surface runoff and percolation, and, in general, to increase or decrease water availability to the aquifer should be carefully controlled. Included are land use pattern, landfilling, lagooning, reservoir construction, deep well injection, and pumping rate. Complete tests should be made to investigate the existing groundwater hydrology and correctional techniques, as they relate to land slope and topography; surface area; reservoir, lagoon, and landfill lining; and deep well injection. Pumping rates may be adjusted to minimize the impact.

Secondary Effects.   Alterations in aquifer safe yield can be related to other attributes in terms of secondary impacts. Aside from being a community need, a safe, dependable water supply is necessary for community and for regional economic stability. It can affect land use patterns as well, since it is a factor in domestic, industrial, and agricultural requirements.

Flow Variations

Definition of the Attribute.   The velocity of flow and discharge are extremely important to aquatic organisms in a number of ways, including the transport of nutrients and organic food past those organisms attached to stationary surfaces; the transport of plankton and benthos as drift, which in turn serve as food for higher organisms; and the addition of oxygen to the water through surface aeration. Silts are moved downstream and sediments may be transported as bed load. These, in turn, are often associated with major nutrients, such as nitrogen and phosphorus, which may be released at some point downstream.

Natural flow variations are, therefore, critical factors governing the type of ecological system that will develop and survive in a given watercourse. If the pattern of stream-flow variation is changed markedly from that which is natural, subsequent disruption of the natural ecology may result.

Activities that Affect the Attribute.   Major activities that may influence stream flow include reservoir projects, and changing the ground surface and topography for different types of land use projects. This may include site clearing, earthwork and borrowing, paving of land areas, and building construction. Other activities include modification of vegetation, which can lead to altered runoff patterns, and water use changes in withdrawal and return flow rates.

Source of Effects.   Reservoir projects may be flood control (that reduces high flows), power generation (that minimizes low flow conditions), or any desired use that alters the flow pattern of the stream. The land use project alters the runoff, percolation, and evaporation in the drainage basin. These changes may increase or decrease the runoff. Other attributes affected by such activities are suspended solids and nutrients in the watercourses; they may, in turn, affect the population of photosynthetic organisms, and thus, the food chain. Direct flow variations are caused by fluctuating municipal, industrial, and/or agricultural demands and the return flows from these users.

Variables to be Measured; How Variables are Measured.   Flow measurement is relatively simple. Many types of automatic flow-measurement devices that can be installed in a selected reach of a watercourse are commercially available. The unit of flow measurement is cubic feet per second (CFS).

Evaluation and Interpretation of Data.   If flow variations are rapid and extensive, more disruption to natural ecology results. However, due to lack of information, classification of water cannot be made on the basis of qualitative measurement. Five degrees of environmental impact are summarized in Table B-4.   This classification is based upon qualitative or observed conditions.

Special Conditions.   Flow variations become most significant at the extreme conditions—low flows and high flows. Under low flow conditions, the natural assimilative capacity of a given stream is greatly reduced, and the adverse effects of natural and man-induced waste loads are most critical. At high flows, physical damage due to flooding and inundation of vegetation becomes of major concern.

Geographical and Temporal Limitations.   Low flow considerations are of importance on all streams; however, they warrant particular consideration in areas which typically experience prolonged periods of drought. These periods frequently coincide with summer, when biological activity is high, and dissolved oxygen content in streams is at a minimum, thus compounding the significance of the problem.
Impacts associated with high flow conditions are most likely to occur in areas and climates with conditions conducive to flooding.

Mitigation of Impact.   All activities such as land use projects and water impoundment and operation should be given consideration to minimize flow variations from the mean natural flow.

Secondary Effects.   Man-induced flow variations may have secondary impacts in ecology, land use patterns, and in the socioeconomic realm. Many species of plant and animal life are sensitive to flow variations and require specific ranges of flow conditions. Floodplain development can be a function of the degree of control over flow variations. Economic losses are felt through flooding of agricultural as well as built-up areas, and adverse psychological effects are apparent when there are threats of flood.

Oil

Definition of the Attribute.   Oil slicks are barely visible at a concentration of about 25 gallons/square mile.   At about 50 gallons per square mile, an oil film is about 3.0×10^{-6} inch thick and is visible as a silvery sheen on the surface.   Oil is destructive to aquatic life in the following ways:

Free oil and emulsions may coat and destroy algae and plankton
Heavy coating may interfere with the natural processes of reaeration and photosynthesis
Water-soluble fractions may exert a direct toxic action
Settleable oil substances may coat the bottom, destroy benthic organisms, and interfere with spawning areas.

Activities that Affect the Attribute.   Major activities responsible for oil pollution include bilge and ballast waters from ships; oil refinery wastes; industrial plant wastes such as oil, grease, and fats, and lubrication of machinery; gasoline filling stations; bulk stations; and accidental spills.

Source of Effects.   Oil may reach natural waters by direct discharge or by surface runoff.   Direct discharge may occur from bilge and ballast waters or by accidental spill from barges or tankers.   Indirect oil release may occur from surface runoff or storm sewers, or combined sewer overflows.   In all cases, damage could be severe and long lasting.   Water quality parameters affected by oil discharge are dissolved oxygen, general appearance, and taste and odor.

Variables to be Measured; How Variables are Measured.   Dissolved or emulsified oil or grease is extracted from water by intimate contact with various organic solvents.   The results are expressed in mg/l oil or grease. Other measurements are qualitative and include (1) visible oil slick, (2) oily taste and odor of fish and edible invertebrates and (3) coating of banks and bottom or tainting of associated biota.   Quantitative measurement of oil and grease is by extraction in a separating funnel with either trichlorotrifluoro-ethane or petroleum ether.   The technique is used as routine analysis in water and wastewater analysis.

Evaluation and Interpretation of Data.   Due to lack of information, classi-fication of water cannot be made on the basis of quantitative measurement or concentrations.   Five degrees of water impacts are summarized in Table B-4 on the basis of qualitative or observed conditions.

Mitigation of Impact.   Oil pollution can be minimized by controlling all direct discharge into natural waters.   Surface runoff from oil handling areas should be treated for oil separation before discharge into the environment. If oil wastes are combined with sanitary sewage, oil separation will be necessary at the wastewater treatment facility. Lagooning of oil wastes and land disposal of oily sludges should be restricted in order to avoid possible contamination of the groundwater system.

Secondary Effects. Secondary effects of oil discharges are manifested through impacts on aquatic ecology and waterfowl, economic loss through decreased recreational desirability, and lowered property values if the discharges become frequent. Increased activity in exploration, production, and transportation of petroleum can increase controversy, divide communities, alter land use patterns, and indirectly affect public and private land markets—whether or not any actual spills take place.

Radioactivity

Definition of the Attribute. Ionizing radiation, when absorbed in living tissue in quantities substantially above that of natural background, is injurious. It is, therefore, necessary to prevent excessive levels of radiation from reaching any organism, be it human, fish, or invertebrate.

Activities that Affect the Attribute. Human activities responsible for radiation hazards are application of nuclear methods in power development, industrial operation, medical laboratories, research and development, nuclear weapon testing, and radiation warfare. In all applications, radioactive substances may be released accidentally, by inadequately planned and controlled activity, or by disposal of radioactive wastes.

Source of Effects. Radioactivity, once released to the aquatic environment, may (1) remain in solution or in suspension, (2) precipitate and settle to the bottom, or (3) be taken up by plants and animals. Immediately upon introduction of radioactive materials into the water, the wastes may become diluted by dispersion or may become concentrated by the process of biological magnification.*

Variables to be Measured; How Variables are Measured. The measure of radioactivity is the curie, the quantity of any radioactive material in which the disintegrations per second are 3.7×10^{10}; this is a large amount of radioactivity. Two smaller units, microcurie (10^{-6} curie) or picocurie (10^{-12} curie or 2.22 disintegrations per minute), are often used. Radioactive waste can be

*This is a process by which some substances become concentrated instead of dispersed with each link in the food chain.

diluted in water to below the allowable limit. The allowable limit of radiation in natural water is taken as 10^{-7} microcurie per ml when the activity is caused by an unknown mixture of beta- and gamma-emitting isotopes.

Measurement techniques are not difficult because radiation-counting equipment of high sensitivity and stability is commercially available.

Evaluation and Interpretation of Data.   It is not easy to determine the long-term effects of the radiological wastes upon aquatic life. For this reason, and as a practical matter, radioactivity exceeding the allowable limit of 10^{-7} microcurie per ml may be considered detrimental to human health and aquatic life. Five classes of water impacts are given in Table B-4.

Special Conditions.   Special precautions should be taken to prevent radioactive materials from entering ground or surface waters to be used for supply, fish production, or recreation.

Mitigation of Impact.   Release of radioactive wastes from radiation facilities must be monitored and controlled. Radioactivity in sewage after treatment is reduced in unknown amounts through concentration in sludge. However, sludge disposal becomes a difficult problem. Therefore, waste containing radioactivity should be treated separately by means of de-watering procedures, and solids or brine should be disposed of by special care (deep well injection or containment). Fallout of radioactive dust will induce radioactivity in surface runoff, the treatment of which is a difficult task. All efforts should, therefore, be made to minimize release of radioactivity into the environment.

Secondary Effects.   While it is generally understood that aquatic organisms are relatively tolerant of radioactive materials, little is known of the mechanism of concentrating radioactive elements by these organisms and the effect this might have on human or other consumer organisms. Whereas there have been few cases of actual radioactive contamination of water resources, the fear of such contamination actually has had much greater impact. These fears have resulted in controversy, altered land use patterns, and have had other socioeconomic effects.

Suspended Solids

Definition of the Attribute.   Suspended solids are solids contained in water which are not in solution. They are distinguished from dissolved solids by laboratory filtration tests. Suspended solids comprise settleable, floating (specific gravity lower than water), and nonsettleable (colloidal suspension) components. These may contain organic (volatile suspended solids) or inert (nonvolatile) substances. Turbidity may be caused by a wide variety of suspended materials, ranging in size from colloidal particles to a coarse dispersion, depending upon the turbulence.

Suspended solids are perhaps of greatest significance from the standpoint of aesthetics. Natural waters may contain wide variations of suspended solids. These may be due to clay, silt, silica, organic matter, microorganisms, or sewage. Suspended solids may be undesirable in many ways. In public water supplies, turbid water is difficult and costly to filter. Disinfection may require higher chemical dosages if the water is turbid. Also, excessive suspended solids can be harmful to fish and other aquatic life by coating gills, blanketing bottom organisms, reducing solar radiation intensity, and, thus, affecting the natural food chain. In stream pollution-control work, all suspended solids are considered to be settleable solids, because eventually (by bacterial decomposition and chemical flocculation) those solids are deposited.

Activities that Affect the Attribute. Activities directly responsible for suspended-solids release are dredging, wastewater discharge, construction of hydraulic structures, and gravel washing. Activities that indirectly affect suspended solids result from land use: site clearing, surface paving, building construction, landscaping, and mine tailings. All change the surface runoff pattern, which, in most cases, increases the storm flow. Suspended-solid load in the surface runoff may change considerably due to erosion. Also, flow variations in streams may change the bed load and solids transport.

Source of Effects. As discussed above, many activities will increase or decrease the suspended-solid condition in natural waters. It may be mentioned that many times this effect may be temporary. For example, dredging may increase suspended solids during operation. After completion of dredging, the channel may become deeper and wider; thus, dredging may actually reduce velocity and encourage settling. Likewise, many other activities, such as construction, site clearing, and excavation may have effects that should be evaluated as long-term or short-term effects.

Many water quality attributes may be affected by change in suspended solid condition. These include DO (due to increase in photosynthesis), nutrient enrichment, and direct deleterious effect to fish and other aquatic life by coating gills or blanketing bottom organisms, for example.

Variables to be Measured; How Variables are Measured. Settleable suspended solids are measured in mg/l of settled water. Suspended solids are measured by filtering a sample through a membrane filter or an asbestos mat in a Gooch crucible. Turbidity is measured in Jackson units equivalent to the interference to light transmission caused by 1 mg/l of a standard suspension.

Many types of commerically available instruments can continuously measure and record the turbidity and the suspended solids in water. They all rely upon passage of light through a standard light path.

Evaluation and Interpretation of Data.   Water quality is considered lower with increasing turbidity and suspended solids. Table B-4 summarizes the five classes of water impact, based upon turbidity, suspended solids, and visual consideration.

Mitigation of Impact.   The impact due to suspended solids may be minimized by controlling discharge of wastes that contain suspended solids; this includes sanitary sewage and industrial wastes. Also, all activity that increases erosion or contributes nutrients to water (thus stimulating algae growth) should be minimized.
The gravel washing activity, mine tailings, and anything causing dust may be controlled by utilizing available technology.

Secondary Effects.   Increase in suspended solids content may have secondary impact on socioeconomic attributes as a result of loss of productivity (e.g., decline in fish harvest) and reduction in various recreation-oriented activities. Additionally, increased costs to remove suspended solids for domestic or industrial water use may occur as a result. Long-term effects include siltation of reservoirs and reducing useful capacity, and filling of marsh areas in estuaries, reducing productive habitat.

Thermal Discharge

Definition of the Attribute.   Temperature is a prime regulator of natural processes within the water environment. It governs physiological function in organisms, and, acting directly or indirectly in combination with other water quality constituents, affects aquatic life with each change. Water temperature controls spawning and hatching, regulates activity, and stimulates or suppresses growth and dvelopment; it can kill when the water becomes heated or chilled too suddenly. Colder water generally suppresses development; warmer water generally accelerates activity.

Activities that Affect the Attribute.   Human activities affecting the attribute are discharges with temperatures above or below that of the receiving waters. Heated discharge may result from sources such as thermal power generation, heavy machine operations, and industrial operations.
Cold water discharges may result from flows from large, deep reservoirs.

Source of Effects.   Heated wastes, when discharged into the water environment, raise the temperature of the water. The extent to which the temperature is raised depends upon the quantity of waste heat discharged and the amount of diluting water available.   As water temperature increases, the solubility of oxygen decreases. Furthermore, the accelerated biological activity imposes higher oxygen demand.   The net result is a decrease in DO level which can reach critical levels.

Water released from lower depths of stratified reservoirs may be significantly lower in temperature and DO content than would prevail in normal ambient stream conditions.   Thus, release depths can have a pronounced effect upon the aquatic life below reservoirs.

Variables to be Measured; How Variables are Measured.   Temperature measurement is simple and accurate.   Many types of automatic temperature recording devices are commercially available.   Measurement scale is either degrees Centigrade or Fahrenheit.   Prediction of the effects of projects on ambient water temperatures is a complex problem which may be addressed through the use of mathematical models for heat exchange in the aquatic environment.

Evaluation and Interpretation of Data.   In environmental quality assessment, the temperature effects are best handled in terms of the magnitude of departure from the natural conditions.   Table B-4 summarizes five classes of water impacts, based upon temperature rise above natural conditions. Allowable departures from ambient temperatures may vary with location, so state water quality regulations should be consulted.

Geographical and Temporal Limitations.   Fogging problems may be associated with warm water discharges in cold regions or under special climatic conditions.

Mitigation of Impact.   Cooling towers can be used to convert once-through systems into closed systems.   A very efficient way is to utilize treated wastewaters (such as sewage, industrial wastes, or stored surface runoffs) as cooling water makeup.   Many industrial plants are considering such a closed system. Chromium may be recovered from cooling tower blowdown before treatment and disposal of tower blowdown.

Secondary Effects.   Effects on the aquatic environment resulting from temperature alteration may, in turn, have other biophysical and socioeconomic consequences.   Increased heat to water bodies accelerates evaporation, and thus, increases the suspended solids content of the water.   This and other impacts on the biological activity may alter the aesthetic and recreational desirability of a given area. Depending upon the circumstance, these effects may be of a positive or negative nature. On the one hand, heat addition may speed the eutrophication process and reduce recreational use. In other instances, this effect has increased recreational benefit through increased productivity. Aquaculture, or "fish farming," has been investigated as a possible beneficial secondary result of heated discharges.

Acid and Alkali

Definition of the Attribute.   Acid and alkaline wastes discharged into waters may change the natural buffer system. pH of the water may significantly change, depending upon the extent of acid or alkali discharged. Change in pH of natural water is hazardous for fish and other aquatic life. Below a pH of 5.0 and above 9.0, fish mortalities may be expected.

Activities that Affect the Attribute.   Activities which may contribute acid and alkali waste to the environment are industrial wastes such as pickle liquors, accidental spills of chemicals, and mining operations.

Source of Effects.   Acid and alkali wastes can be extremely damaging to aquatic life. Toxicity due to the presence of heavy metals is increased by synergism. Also, the capacity of natural waters to assimilate organic wastes is significantly reduced by these wastes.

Variables to be Measured; How Variables are Measured.   pH is considered to be an important measure of environment quality. High pH reflects an alkaline situation and low pH reflects an acid condition (a neutral solution has a pH equal to 7.0).
   pH measurement is simple. Many types of continuous measuring and recording instruments are commercially available for this purpose.

Evaluation and Interpretation of Data.   Since the natural pH of aquatic ecosystems varies from one locale to another, the best measure of pH is in terms of departure from natural levels. Table B-4 summarizes five classes of water impacts, based upon pH departure from the normal. It has been assumed that both positive and negative departures are equally damaging to the water environment. This may not be strictly true in normal cases, but due to lack of evidence, such assumptions may be considered valid.

Special Conditions.   In some cases, alkaline or acid wastes actually may help to balance a pH problem. Acid mine drainage is an example of a problem which would be neutralized by an alkaline discharge.

Secondary Effects.   Secondary effects of impacts on the acidity or alkalinity of waters follow as a result of any condition that deteriorates the quality of water. Social and economic losses, in terms of reduced productivity, decline in recreational benefits, and additional costs of treatment to correct problems related to pH are a few examples.

Biochemical Oxygen Demand (BOD)

Definition of the Attribute.   BOD of water is an indirect measure of the amount of biologically degradable organic material present. It is, thus, an indication of the amount of dissolved oxygen (DO) that will be depleted from water during the natural biological assimilation of organic pollutants. The BOD test is widely used to determine the pollutional strength of sewage and industrial wastes in terms of oxygen that would be required if these wastes were discharged into natural waters in which aerobic conditions exist. The test is one of the most important in stream pollution-control activities. By its use, it is possible to determine the degree of pollution in natural waters at any time. This test is also of prime importance in regulatory work and in studies designed to evaluate purification capacity of receiving bodies of water.

Activities that Affect the Attribute.   Activites associated with normal operations, maintenance, and repair may contribute to BOD wastes. These human activities, e.g., sanitary sewage, wastewaters from hospitals, food-handling establishments, laundry facilities, and floor washing from shops constitute BOD wastes. If all wastes are collected by a network of sewers to a central location, adequate treatment must be provided to minimize impact upon the surface-water system. If cesspools, septic tanks, and soakpits are utilized, groundwater in the vicinity may become adversely affected.

Source of Effects.   The discharge of wastes containing organic material imposes oxygen demand in the natural body of water and reduces the DO level. If wastewaters are treated, the combined sewer overflows and surface runoffs may also exert effects under wet weather conditions. All parameters directly or indirectly related to DO also affect the organic waste assimilation. These parameters include depth of water, velocity of flow, temperature, and wind velocity (see secton on DO for general discussion).

Variables to be Measured; How Variables are Measured.   BOD values are generally expressed as the amount of oxygen consumed (mg/l) by organisms during a 5 day period at 20° C. Several other parameters, such as Chemical Oxygen Demand (COD) and Total Organic Carbon (TOC), are also used to represent the organic matter in water and wastewater. COD value indicates the total amount of oxidizable material present and includes BOD. TOC is a measure of bound carbon. Both these tests are closely related to BOD and are used in water and wastewater monitoring programs.

Routine BOD measurements are made in laboratories by dilution techniques; results are obtained in five days. Some modifications of BOD tests may require less time. COD and TOC measurements take only a few hours. Several types of instruments are commercially available which measure TOC more or less on a continuous basis.

Evaluation and Interpretation of Data.   Table B-4 indicates five classes of water: very clean, clean, fairly clean, doubtful, and bad, depending upon the BOD of water. It may be mentioned, however, that this classification must be used on relative terms. As an example, a sluggish stream, reservoir, or lake may show undesirable conditions at BOD of 5 mg/l, whereas a swift mountain stream may easily handle 50 mg/l of BOD without significant deleterious effects.

Mitigation of Impacts.   All wastes containing organic wastes should be processed by treatment methods. The treatment methods may include biological or chemical processes. Also, several types of packaged treatment units are commercially available that can be installed for desired applications.

Secondary Effects.   By virtue of the biologic and aesthetic effects of BOD on aquatic environments, secondary impacts are manifested in terms of additional impacts on aesthetics, reduced recreational benefits, and costs to alleviate the direct consequences of BOD on waters scheduled for reuse. The success of land use planning efforts in areas where water is an integral part of the planning effort (e.g., recreational areas, industrial siting, etc.) is dependent upon the quality of those waters. BOD is a parameter of utmost importance.

Dissolved Oxygen (DO)

Definition of the Attribute.   All living organisms depend upon oxygen in one form or another for their metabolic process. Aerobic organisms require DO and produce innocuous end products. Anaerobic organisms utilize chemically bound oxygen, such as that from sulfates, nitrates, and phosphates, and the end products are odorous. For a diversified warm-water biota, including game fish, DO concentration should remain above 5 mg/l. Absence of DO will lead to the development of anaerobic conditions with odor and aesthetic problems. In surface waters, DO is measured frequently to maintain conditions favorable for the growth and reproduction of fish and other favorable aquatic life.

Activities that Affect the Attribute.   The activities discussed in BOD also apply to DO.   Other activities that may influence DO include site preparation demolition, dredging, and excavation, all of which may cause turbidity and nutrient release.   Routine operations, such as operation and maintenance of aircraft, watercraft, and automotive equipment, may cause oil release.   Oil film interferes with the natural process of reaeration.

Source of Effects.   Discharge of all organic wastes will lower the DO in receiving waters.   A shallow and swift mountain stream can assimilate large quantities of organic wastes without deleterious effects.   This is because swift-moving streams have greater capacity for natural reaeration and for preventing deposition of organic materials at the stream bed.   In a sluggish stream or reservoir, small amounts of BOD released may cause relatively large adverse effects.   The solubility of oxygen in water decreases with increases in temperature and dissolved salts (in freshwater, solubility of oxygen at $0°$ C is 14.6 mg/l, and at $35°$ C, it is 7 mg/l).   Biological activity is also increased at higher temperatures, and thus, the rate of DO utilization from natural waters is significantly increased.   Therefore, BOD wastes discharged into natural waters have more pronounced effects during summer months, when the water is warm.   Thus, water quality parameters, such as temperature, dissolved salts, depth and velocity of stream, wind velocity, and natural reaeration, are all interdependent.   Also in nutrient rich bodies of water, due to algae bloom, the DO level may reach supersaturation during sunny days.   At night, however, the DO level drops considerably, due to lack of photosynthesis.   High turbidity in water may also interfere with photosynthesis by reducing the depth of light penetration.   Oil slicks may reduce the natural reaeration process, too.   Therefore, nutrients, algae, sunny days, turbidity, and oil slicks are all interdependent parameters.

Variables to be Measured; How Variables are Measured.   The unit of DO measurement is mg/l.   It can be measured by titration techniques using the Azide Modification method.   Many commercially available DO meters can be used for DO measurement.

Evaluation and Interpretation of Data.   The oxygen requirements for fish vary with species and age.   Cold water fish require higher oxygen concentration than do the coarse fish (carp, pike, eel).   It may be stated that the 3 to 6 mg/l range is the critical level of DO for nearly all fish.   Below 3 mg/l, further decrease in DO is important only insofar as the development of local anaerobic conditions are concerned; the major damage to fish and aquatic life will already have occurred.   Above 6 mg/l, the major advantage of additional DO is as a reservoir or buffer to handle shock loads of high oxygen demanding waste loads. Table B-4 indicates five classes of water according to DO levels.

Geographical and Temporal Limitations. Typically, the most critical DO problems occur in summer, when biological activity is high and saturated DO content is low.

Mitigation of Impact. The methods are the same as those given for BOD.

Secondary Effects. Secondary impacts are the same as those listed for BOD.

Dissolved Solids

Definition of the Attribute High amounts of total dissolved solids (TDS) are objectionable because of physiological effects, mineral tastes, or economic effects. TDS is the aggregate of carbonates, bicarbonates, chlorides, sulfates, phosphates, nitrates, and other salts of calcium, magnesium, sodium, potassium, and other substances. All salts in solution change the physical and chemical nature of the water and exert osmotic pressure; the magnitude of the change is, to a large extent, dependent upon the total salt concentration (salinity).

Activities that Affect the Attribute. Major areas which may contribute to TDS include mining and quarrying, municipal and industrial waste disposal, brine disposal, lagooning, landfilling of solid wastes, and accidental spill of chemicals.

Source of Effects. Major activities listed above may cause release of salts either directly or indirectly into the natural water system. Direct release includes discharging the waste laden with salts into the water system. Indirect release may be due to runoff from the affected land, or seepage from filled areas. Landfill seepage or leaching may affect groundwater quality, and, if groundwater feeds the water courses, surface water may be affected, as well.

As a result of salt discharge, many water quality parameters will be affected. DO will decrease as a result of high salinity. High quantities of salts give mineral taste. Sulfates and chlorides are associated with corrosion damage. Sulfate in water has a laxative effect. Nitrate plus nitrite causes methomoglobinemia (blue baby disease). Water containing high TDS also exhibits hardness.

Variables to be Measured; How Variables are Measured. Total dissolved solids is determined after evaporation of a sample of water and its subsequent drying in an oven at a definite temperature. This includes "nonfilterable residue." The results are expressed in mg/l TDS.

Evaluation of Interpretation of Data.   For reasons of palatability and unfavorable physiological reaction, a limit of 500 mg/l TDS in potable water has been recommended. Highly mineralized waters are also unsuitable for many industrial applications. Irrigation crops are highly sensitive to salt concentrations; waters containing over 2000 mg/l are of marginal value for irrigation use, and waters containing 3000 mg/l are unsuitable. The upper limits for some freshwater fish are as high as 5000 mg/l. In such cases, reference is only to total salt concentration and its effects on osmotic pressure. Based upon TDS, the five impact classes are summarized in Table B-4.

Special Conditions.   The amount of dissolved ionic matter in a sample may often be estimated by multiplying the specific conductance by an empirical factor. After the empirical factor is established, for a comparatively constant water quality, specific-conductance measurement will yield TDS. Specific-conductance measurement is relatively simple and is a measure of a water's capacity to convey an electric current at 25° C. Specific conductance is expressed as microohms/cm.

Mitigation of Impact.   Wastes containing high TDS are difficult to treat. Recommended treatment methods include removal of liquid and disposal of residue by controlled landfilling to avoid any possible leaching of the fills. Deep well injection has been used for disposal of brine. All surface runoffs around mines or quarries should be collected and concentrated. The brine may be disposed of by deep well injection or other means acceptable to water quality control authorities.

Secondary Effects.   Effects on irrigated crop land (reduced productivity and economic loss) constitute perhaps the most significant secondary impacts due to TDS. Other effects include those on health, where drinking waters are concerned; and those on economics and land use, where industrial and municipal consumption are to be considered.

Nutrients

Definition of the Attribute.   Eutrophication is a term meaning enrichment of waters by nutrients through either man-made or natural means. Present knowledge indicates that fertilizing elements most responsible for eutrophication are phosphorus and nitrogen. Inorganic carbon, iron, and certain trace elements are also important. Eutrophication results in an increase in algae and weed nuisances and an increase in larvae and adult insects. Dense algae growths may form surface water scums and algae-littered beaches. Water may become foul smelling when algae cells die; oxygen is used in decomposition, and fish kills result. Filter-clogging problems at municipal water-treatment plants and taste and odor in water supplies may all be due to dense algae population.

Activites that Affect the Attribute.   Sewage and sewage effluent contain a generous amount of the nutrients necessary for eutrophication. Treated or untreated sewage discharge will contribute to nutrients in receiving waters. Mining, tunneling, blasting, and quarrying into phosphate rocks may cause increased phosphorous from surface runoff. Dredging of waterways will release the storehouse of nutrients contained within the mud bottom; as a result, the water will become enriched during and soon after the dredging operation. Many other activities may also enrich the natural waters. These include drainage from cultivated agricultural lands, surface irrigation returns, dead trees and leaves, logging and sawmilling, and dead organisms.

Source of Effects.   Nutrients released from many activities (described above) will cause aquatic plant problems, turbidity, taste, and odor; cause reservoir and other standing waters to collect nutrients and to store a portion of these within consolidated sediments (once nutrients are combined within the ecosystem of receiving waters, their removal by natural process is very slow); and induce excessive weed growth, which will eventually block waterways or turn lakes into swamps.

As a result of nutrients released into natural waters, many water quality parameters will be affected directly or indirectly. Some of these effects are: turbidity, due to excessive algae growth—then, when algae cells and other plants die, oxygen is used in decomposition and the DOD level declines, causing fish kill; rapid decomposition of dense algae scums, giving rise to odors and hydrogen sulfide gas that create strong citizen disapproval; and serious water-treatment problems, caused by color, taste, and odor.

Variables to be Measured; How Variables are Measured.   Phosphorus, nitrogen, carbon, iron, and trace metals all act as nutrients. Growth of aquatic plants is governed by the law of minimum; i.e., any nutrient, out of a broad array of materials required for growth and development, governs the growth if it is present in a limiting concentration. Most commonly, in natural waters, phosphorus is present in limiting amounts and governs the normal plant growth.

Phosphorus occurs in natural waters and in wastewaters almost solely in the form of phosphates. These forms are commonly classified into orthophosphates, condensed phosphates (pyro-, meta-, and polyphosphates), and organically bound phosphates. These phosphates may occur in the soluble form, in particles of detritus, or in the bodies of aquatic organisms. Because the ratio of total phosphorus to that form of phosphorus readily available for plant growth is constantly changing and ranges from 2 to 17 times or greater, it is desirable to establish limits in the total phosphorus, rather than the portion that may be available for immediate plant use.

Phosphate analysis embodies two general procedural steps: (a) conversion of the phosphorus form of interest to soluble orthophosphate, and (b) colorimetric determination of soluble orthophosphates. The result may be expressed as mg/1 P (phosphorus).

Evaluation and Interpretation of Data.   Although the concentration of inorganic phosphorus that will produce problems varies with the nature of the aquatic environment and the levels of other nutrients, most relatively uncontaminated lake districts are known to have surface waters that contain 0.001 to 0.003 mg/1 total phosphorus as P (they are nutrient deficient). Above 0.02 mg/1 P, one gets into a region of potential algae bloom. Above 0.1 mg/1 P, water is excessively enriched. Table B-4 categorizes five classes of waters, based upon total P contact.

Geographical and Temporal Limitations.   Since algae growth is temperature-dependant, adverse effects due to eutrophication in northern climates are more pronounced in summers. In southern climates, the effects are felt over the entire year; again, with summer the most critical season.

Mitigation of Impact.   Once nutrients are combined within the ecosystem of the receiving waters, their removal is tedious and expensive. In a lake, reservoir, or pond, phosphorus is removed naturally only by overflow, by insects that hatch and fly out of drainage basins, by harvesting a crop (such as fish) and by combination with consolidated bottom sediments.

The most desirable method to mitigate impact is to treat wastewater to a desired phosphorus level before discharge into the environment. Also, all activities mentioned above should be performed under controlled conditions.

Secondary Effects.   Various adverse secondary impacts occur with advanced stages of eutrophication, including a decline in recreational benefits, effects on land use, and the economic losses that normally accompany any deterioration in water quality.

Toxic Compounds

Definition of the Attribute.   Wastes containing concentrations of heavy metals (mercury, copper, silver, lead, nickel, cobalt, arsenic, cadmium, chromium, etc.), either individually or in combination, may be toxic to aquatic organisms, and thus, have a severe impact on the water community.

Other toxic substances include pesticides, ammonia-ammonium compounds, cyanides, sulfides, fluorides, and petrochemical wastes. A severely toxic substance will eliminate aquatic biota until dilution, dissipation, or volatilization reduces concentration below the toxic threshold. Less generally, toxic materials will reduce the aquatic biota, except those species that are able to tolerate the observed concentration of the toxicant. Because toxic materials offer no increased food supply, such as discussed for organic wastes, there is no sharp increase in the population of those organisms that may tolerate a specific concentration.

Activities that Affect the Attribute.   Many human activities may contribute to release of toxic compounds into the environment. These include waste discharged from maintenance and repair shops, and from industrial operations. Wastes that are particularly likely to contain toxic compounds result from electroplating, galvanizing, metal finishing, and cooling tower blow down. Other activities which may contribute to toxic chemicals are mining, accidental spills of chemicals, chemical warfare, and leaching of landfills containing toxic compounds.

Source of Effects.   Chemicals released into the environment may effect surface water or groundwater systems by direct discharge of wastes containing toxic compounds or from surface runoff which may come in contact with toxic material left as residue over the ground surface.

Variables to be Measured; How Variables are Measured.   The spectrum of toxic materials is extremely large and highly diverse in terms of effects. Measurement may be expressed as mg/l of specific compound under consideration. For a group of toxic compounds, it should be pointed out that possible synergistic or antagonistic interactions between mixed compounds may cause different effects than those associated with the respective toxic compounds considered separately.

Bioassay is an important tool in the investigation of these wastes, because results from such a study indicate that degree of hazard to aquatic life of particular discharges; interpretations and recommendations can be made from these studies concerning the level of discharge that can be tolerated by the receiving aquatic community.

The basic bioassay shall consist of a 96 hour exposure of an appropriate organism, in numbers adequate to assure statistical validity, to an array of concentrations of the substance, or mixture of substances, that will reveal the level of pollution that will cause (1) irreversible damage to 50 percent of the test organisms and (2) the maximum concentration causing no apparent effect on the test organisms in 96 hours.

Evaluation and Interpretation of Data. The bioassay may indicate the concentration at which toxic compounds will not cause an apparent effect upon the test organism. However, long-term effects of toxic compounds having more subtle changes, such as reduced growth, lowered fertility, altered physiology, and induced abnormal patterns, may have more disasterous effects on the continued existence of a species. Also, the biological magnification and storage of toxic residue of polluting substances and microorganisms may have another serious aftereffect. For all these reasons, and as a practical matter, toxic compounds, if they could be detected in natural waters by modern water quality analysis methods, may render water undesirable for propagation of healthy aquatic life. The five classes of water, based upon toxic compounds, are given in Table B-4.

Special Conditions. Synergistic action may magnify toxic effects under special conditions (e.g., under an increased temperature or a low dissolved oxygen situation).

Mitigation of Impact. All wastes containing toxic chemicals should be monitored and controlled. Those released into sanitary sewers should be carefully regulated so that such release does not affect the treatment process. Also, after dilution, effluent concentration should not exceed the desired level. Runoffs from chemical handling areas should also be considered, to the extent that pollution is expected. If necessary, suitable treatment may be given to all contaminated runoffs.

Secondary Effects. While toxic compounds have a primary effect on lower organisms in the aquatic environment, secondary effects may be felt all through the food chain, with human health as a final major consideration. Procedures to remove these compounds, once they have been released to the aquatic environment, may be nonexistent or, at best, extremely expensive. Failure to remove them or to prevent their initial entry may degrade the water quality with the ensuing effects on aesthetics, economics, and biophysical relationships.

Aquatic Life

Definition of the Attribute. Organisms in any community exist in a dynamic state of balance, in which the population of each species is constantly striving to increase. However, population is maintained at a fluctuating level determined by food supply, predators, chemical characteristics of the water, and physical variables. Since these factors vary greatly, several types of communities exist in balance. Any man-made pollution tends to upset the natural state of balance. This may cause abundance of a few types of organisms, while others may decline or completely disappear. Because of some variation in response among species to conditions of existence within the environment, and because of inherent difficulties in aquatic invertebrate taxonomy, ecological evaluation of the total organism community is the acceptable approach in water pollution-control investigation. Today's investigators tend to place organisms in broad groups, according to the general group response to pollutants in the environment.

Activities that Affect the Attribute.   All activities discussed above (with various water attributes) affect aquatic life to some degree. Change in an aquatic community depends upon the type and extent of pollution.

Source of Effects.   Discharge of organic wastes (sewage) tends to lower the natural DO and to eliminate DO-sensitive organisms. Thermal discharge affects the normal life cycle of many organisms. Toxic wastes will reduce the aquatic biota, except those species that are able to tolerate the observed concentration of the toxicant. In general, changes in any attributes, whether they are physical or chemical, will influence the aquatic life.

Variables to be Measured; How Variables are Measured.   For aquatic life interpretation, field observations are indispensable. However, many of the biological parameters cannot be evaluated directly in the field. The specific nature of a problem and the reasons for collecting samples will dictate those aquatic communities of organisms to be examined and those, in turn, will establish sampling and analytical techiques. The following communities and types of organisms are considered: plankton, periphyton, macroinvertebrates, macrophytes, and fish. Sampling and identification techniques are based upon routine biological sampling and analysis methods. Readers are referred to *Standard Methods for the Examination of Water and Wastewater*, 13th Edition, 1971.

Evaluation and Interpretation of Data.   Based upon most common aquatic life in natural waters, five classes of water are given in Table A-4.

Mitigation of Impact.   See all water quality attributes for mitigation of impact upon aquatic life.

Secondary Effects.   Economic and recreational benefits may be affected as a result of adverse impacts on aquatic life. Loss of productivity reduces fishing harvest, and decline in recreational activity produces additional economic loss.

Fecal Coliforms

Definition of the Attribute.   Water acts as a vehicle for the spread of disease. All sewage-contaminated waters must be presumed potentially dangerous. The presence of coliform organisms in water is regarded as evidence of fecal contamination, as their origin is in the intestinal tract of humans and other warm-blooded animals. They are also found in soil and water which has been subjected to pollution by dust, insects, birds, and small and large animals. The necessity of coliform tests in water supply has declined somewhat since water treatment plants effectively remove most of the bacteria by treatment and disinfection.   However, the test continues to retain importance because of water-contact recreational usage of water, and of implications that viral diseases can be transmitted through fecal contamination of water supplies.   Indirect routes, such as the contamination of foods with fecally contaminated irrigation water and accumulation of contaminants by oysters, clams, and mussels from fecally-contaminated marine waters, continue to be areas of concern.

Activities that Affect the Attribute.   The activities discussed in *BOD* and *DO* also apply to this attribute.

Source of Effects.   See *BOD* and *DOD* attributes.

Variables to be Measured; How Variables are Measured.   Two methods are used for determining the presence of coliform organisms: the multiple tube fermentation technique and the membrane filter technique. The results of multiple tube fermentation techniques are expressed as Most Probable Number (MPN), based upon certain probability formulas. The results of membrane filter tests are obtained by actual count of coliform colonies developed over membrane filter. In both cases, the estimated coliform density is reported in terms of coliform per 100 ml. The equipment used are the type commonly needed in routine microbiological study.

Evaluation and Interpretation of Data.   Present water quality criteria restrict the use of water, depending upon fecal coliform density. The desirable criteria for surface water supply is fecal coliform less than 20 per 100 ml, and for recreational use (including primary contact recreation), the recommended value is 200 per 100 ml. Based upon the coliform density, five classes of water are summarized in Table B-4.

Mitigation of Impact.   See attributes *BOD* and *DOD*.

Secondary Effect. Quantification of the presence of fecal coliforms in recreational waters results in a classification by permissible use. This classification restricts not only the use of the waters, but also the economic benefits which might be obtained from those waters. Effects on shellfish harvests are other economic impacts which may result from fecal contamination.

REFERENCES

1. Mackenthun, K. M., *The Practice of Water Pollution Biology*, U.S. Department of the Interior, FWPCA, Division of Technical Reports, 1969.

2. Jain, R. K. *et al.*, *Environmental Impact Study for Army Military Programs*, U.S. Army Construction Engineering Research Lab, Champaign, Ilinois, December, 1973.

3. Odum, E. P., *Fundamentals of Ecology*, W. B. Saunders Co., Philadelphia, Pennsylvania, 1971.

4. McGauhey, P. H., *Engineering Management of Water Quality*, McGraw-Hill Book Co., New York, 1968.

5. Keup, L. E. et. al., *Biology of Water Pollution*, U. S. Dept. of the Interior, FWPCA, Cincinnati, Ohio, 1967.

6. Nemerow, N. L., *Scientific Stream Pollution Analysis*, Scripta Book Co., Washington, D. C., 1974

LAND

As with all other resources available to man, land is not available in unlimited quantities. Because of this, it is becoming increasingly recognized in this country, and in other countries with less of an endowment of land resources, that land use must be properly planned and controlled. CEQ guidelines recognize this need for the rational management of land resources, and, because the price system does not allow rational allocation of land, CEQ has provided for a specific consideration of the relationship of a changed pattern in land use to the existing pattern. Therefore, land is being treated much in the same manner as our other scarce natural resources, air and water.

To consider these factors requires comprehensive consideration of existing and projected land capabilities and land use patterns. The most significant element of the land use question has been collapsed into three attributes:

Erosion
Natural hazards
Land use patterns.

Erosion

Definition of the Attribute.   Erosion is defined as the process through which soil particles are dislodged and transported to other locations by the actions of water and/or wind. The two most common forms attributable to water are sheet erosion, in which the upper surface of the soil is more or less evenly displaced, and gully or rill erosion, in which the downward cutting action of the overland flow of water results in linear excavations deep into the soil horizon. While the latter type of erosion is often more spectacular to the eye, loss of uniform layers of topsoil through sheet erosion is the more serious of the two. Wind erosion is similar to sheet erosion in that very small soil particles containing plant nutrients and organic matter are the ones that are carried away, leaving coarse and less productive material.

Soils of almost all types are held in place on slopes by vegetative cover and its associated root system. Removal of this cover exposes the soil to the erosive forces of water and wind. Erosion is intensely destructive. First, the site itself may be denuded of its most productive topsoils and/or may be gullied to the extent that it becomes almost totally unproductive, often to the point of posing a physical barrier to other activities. Second, the streams and lakes which receive the attendant sediment loads may be affected. The landscape, after erosive forces have been at work, is barren and aesthetically unappealing.

Activities that Affect the Attribute.   Activities that affect the extent and rate of erosion are those associated in any way with removal or re-establishment of vegetative cover. Some of these are land clearing for construction, road building or other cut and fill operations, timber harvesting or vegetative suppression by herbicide application, controlled burning, reforestation or afforestation, strip mining, agricultural activities, off-road vehicular traffic, and large animal grazing.

Source of Effects.   Land clearing and mechanized off-road activities strip land of its vegetative cover, organic surface material, and root structures which formerly protected the soil, thereby opening it up to direct attack by wind and water. Timber harvesting, application of herbicides, and controlled burning can result in the removal of a sufficient quantity of organic surface material and vegetative cover to cause an increase in the intensity of rainfall and wind movement at the soil surface. Conversely, reforestation and afforestation can reintroduce a vegetative canopy and root structure which—over time—can reduce the intensity of these erosive forces and result in a buildup of organic surface material. Road building and other cut-and-fill activities lay bare previously vegetated soil, alter natural drainage patterns, change the gradient of slopes, and create somewhat unconsolidated fill areas upon which vegetative cover is often not immediately reestablished. The stripping away of vegetative shrub and ground cover in semiarid areas by overgrazing is one of the most widespread causes of wind erosion. If grazing rights are not renewed, or large wild animals are fenced out of and/or removed from overgrazed areas, seeding of native grasses can accelerate the return of vegetative cover and reduce erosion potential.

Variables to be Measured. Major variables affecting erosion are soil composition or texture, degree of slope, uninterrupted length of slope, nature and extent of vegetative cover, and intensity and frequency of exposure to the eroding forces. The interaction of these variables is complex, and difficult to measure directly. Magnitude of the impact is also directly dependent on the extent of the affected area.

Soil texture is determined by the percentage of its sand, silt, and clay components. Generally accepted textural classes in order of decreasing particle size (coarse to fine) are:

Sand	Silt loam	Silty clay loam
Loamy sand	Silt	Sandy clay
Sandy loam	Sandy clay loam	Silty clay
Loam	Clay loam	Clay

While such a statement is subject to contradiction on a specific site, finer textured soils are usually more susceptible to water erosion. Sandy soils and granulated clays are those most easily eroded by wind.

Erosion increases with the length and steepness of slope. A general rule is that if the length of slope is doubled, soil loss from erosion will increase by a factor of 1.5. The relationship between degree of slope (gradient)* and erosion potential can be specified in general terms as follows:

$$10 \text{ percent} \geqslant \text{ highly erodible}$$
$$2–10 \text{ percent} = \text{ moderately erodible}$$
$$2 \text{ percent} \leqslant \text{ slightly erodible.}[1]$$

The erosion hazard depends upon the intensity and frequency of rain and wind storms. While the amount of yearly rainfall is important, of greater significance is the force with which it strikes the ground, volume in a given time, and return frequency of intense storms. The impact of wind varies with velocity, direction, and soil moisture content.

The difference in types of vegetative cover and the extent of each also affect erosion potential. A mature forest with a heavy overstory (leaf or needle) cover, an understory of trees with less dense leaves, scattered ground vegeta-

---

*Slope gradient is the relationship between the vertical height and the horizontal length of the slope.

tion, and a heavy layer of decaying organic matter will protect the soil from wind and water to a greater extent than will brush and sparse ground cover found in arid and semiarid areas. These are extremes—pasture and cultivated cropland fall somewhere between.

Before proceeding further, some informed judgment should be made as to whether these variables are operative to a degree and in sufficient combination to warrant the rather extensive calculations to be described next. If necessary, an agronomist or agricultural engineer from the local office of the U. S. Soil Conservation Service (SCS) could assist in making this initial assessment.

How Variables are Measured. Most soil loss or soil erosion equations are based upon models that represent interrelationships among the variables just discussed. One such model developed for agricultural cropland—but subject to modification for other vegetative types is:[2]

$$A = RKLSCP$$

where

A = Computed soil loss per unit area (acre)
R = Rainfall factor
K = Soil erodibility factor
L = Slope-length factor
S = Slope gradient factor
C = Crop management factor
P = Erosion control practice factor.

While the techniques of arriving at numbers to represent the various factors are adequately described in a handbook which should be available from the local office of the Soil Conservation Service (SCS), it would be helpful to have the expert advice of a team of SCS agronomists, hydrologists, and agricultural engineers in applying it to a specific site. Soil loss should be computed both with and without the project.

The area affected should be outlined on a map overlay of appropriate scale; through the use of a planimeter and with the assistance of an engineer, the number of acres affected can be determined. Total soil loss with and without the project can then be calculated by multiplying the soil loss per acre, as previously obtained from the model, by the number of acres involved. This change, expressed as a percentage, can be obtained from the following equation:

$$\text{Percent change in soil loss} = 100 - \left(\frac{\text{Soil loss without project}}{\text{Soil loss with project}}\right) \times 100$$

Evaluation and Interpretation of Data.   Overall magnitude of the impact can be represented by the percent change in total soil loss as calculated above. If a more sophisticated analysis appears to be warranted, this quantitative figure can be tempered by a further evaluation that takes into account change in soil fertility (productive capacity) and the impact of changes in sediment load in streams that drain the affected area. This kind of analysis could best be done by an interdisciplinary team of economists, agronomists, engineers, and ecologists.

Special Conditions.   If the land were productive for agricultural crops or forest products, the economic and ecological impacts might be greater than if it were relatively infertile.

Geographical and Temporal Limitations.   While there are few areas in the United States where the potential for at least moderate erosion does not exist, most severe erosion has occurred in the Appalachian area of the Southeast, in the Great Plains, and in some desert and semiarid areas of the Southwest. The major temporal limitation on erosion involves the time of year when the soil is exposed and the length of time it remains exposed relative to the time of year that intense rain and windstorms are likely to occur.

Mitigation of Impact.   It is much easier to prevent erosion before it begins than it is to arrest it or restore the land afterwards. The environmental impact of soil erosion can best be mitigated by removing vegetative cover only from the specific site on which construction is to take place and by disturbing the vegetation in adjacent areas as little as possible. Construction, land management, or mining activities that result in the soil being laid bare could be scheduled in such a way that some type of vegetative cover appropriate to the site could be established prior to the onset of intense rain or windstorms. If grass is to be seeded, a mulch of straw will help to protect the soil from less extreme erosive forces until vegetative and root development begin. Natural drainage patterns can often be maintained by preparing sodded waterways or installing culverts. Steep slopes can be terraced, thereby effectively reducing the length of slope. Catch basins built near construction sites can reduce the quantity of eroded soil particles reaching free-flowing streams or lakes. Additional information on mitigation techniques is available in a U. S. Environmental Protection Agency report.[3]

Secondary Effects.   Secondary effects of erosion include increased sediment loads in streams which may clog reservoirs and fill large areas of bays and estuaries. These sediment loads also affect aquatic life through such mechanisms as the covering of fish eggs and spawning areas, coating of gills, and retarding light penetration, which, in turn, reduces the photosynthetic process necessary for aquatic plant production. As the aquatic environment is degraded, the results are losses in areas of aesthetic, recreational, and economic benefits. Other economic losses include adverse effects on land use suitability, crop production reduction, and frequent filter replacement due to increased particulate materials in the air.

Other Comments.   If the erosive effects resulting from an activity are not confined but spill over into adjacent private lands (sediment deposition), or if severely eroded land is visible from public highways, after-the-fact controversy over the project may develop. This is especially true if these considerations are not directly addressed in the environmental assessment/impact statement and if the mitigation possibilities are not discussed and evaluated.

Natural Hazards

Definition of the Attribute.   Natural hazards are those occurrences brought about by the forces of nature that cause discomfort, injury, or death to man; damage or destroy physical structures and other real or personal property; change the physical character of land, water, and air; and damage or destroy the plant and animal life of the affected area. The severity and frequency of occurrence of floods, earthslides, and wildfires may be influenced by various activities. Other natural hazards, such as earthquakes and hurricanes, may cause greater personal and physical damage than would be the case if human activities were located in areas other than those where these natural events occur with some frequency and severity.

Activities that Affect the Attribute.   Some activities that often have an impact on the frequency and magnitude of natural hazards are construction, land management, land use, agriculture, and industrial development. These activities do not affect the natural processes that are the root causes of hazards—intense rain or wind storms, the geologic structure and soil and bedrock properties of an area, or lightning strikes from the thunderstorms. Rather, it is the destructive nature of the results of these occurrences that human activities influence.

Source of Effects.   The effects of construction activities on the destructive potential of natural hazards are quite diverse. Land clearing, which precedes most kinds of construction, lays bare the soil surface, a condition conducive to increased volumes of water runoff and increased sediment loads in streams—both of which tend to cause increases in flood heights and return frequencies, the two greatest determinants of flood damage. Paving large areas with asphalt and concrete—often done for vehicle parks and outdoor storage areas—reduces infiltration of water into the soil, thereby increasing runoff and the peak volume of water that streams are required to carry. The building of structures such as dams and levies, as well as stream channalization to reduce flood levels, may greatly modify the flow regimes of natural water courses, which, in turn, may result in the diversion of flood waters to previously unflooded areas.

The probable incidence of earthslides may be increased by road construction activities if natural shear stresses in the earth are increased, excessive pore pressure developed, or rock and soil strata exposed by road cuts. Failure may be induced by blasting, changes in slope, greater overburden, etc. Earthslides can block streams and cause a backup of water, which, in turn, can result in upstream damage due to a gradual rise in water level and extensive downstream damage due to the rapid release of water when the slide is overtopped or eaten away. Earthslides also destroy vegetation, increase sediment loads in streams, and disrupt transportation routes. On the positive side, road construction in remote areas can reduce potential wildfire damage by permitting more rapid access by fire fighting crews and equipment.

Land management includes activities such as timber harvest, reforestation and afforestation, herbicide application, and controlled burning. Timber harvest can create at least temporary increases in runoff volume and sediment loads as a result of the removal of some of the vegetation cover and the disturbance of the soil surface by trucks and other mechanized equipment. Rehabilitation of eroded areas by reforestation, afforestation, or seeding decreases runoff and sediment loads. Timber harvest on steep slopes can result in landslides which disturb the soil horizon to the extent that natural tree regeneration will not take place. When vegetation killed by herbicides and logging debris left after timber-harvest operations dries to the point where the plant material will ignite easily and burn with considerable intensity, lightning strikes are more likely to cause fires that are difficult to control and may do great damage. Conversely, controlled burning can reduce the incidence and destructive potential of wildfire by creating a low-temperature blaze that consumes the dry underbrush and organic matter on the forest floor without damaging mature timber. (This favorable impact of controlled burning should not overshadow the fact that it may adversely affect other environmental attributes, such as vegetative diversity, wildlife populations, and erosion.)

Land use considerations that dictate where certain projects will be located often have a decided impact on natural hazards. Any physical structure (building, bridge pier, or temporary bridge) that occupies a portion of the floodway (the stream channel carrying the normal water flow) or is situated on the floodplain (that area covered by flood waters when a stream overflows its banks) will restrict the flow of water and decrease the volume which the floodplain can accommodate at a particular level, thereby increasing flood heights both upstream and downstream. Permitting homes or other structures to be located in floodplains poses the possibility of increased physical damage to the structures and loss of life to their occupants. During a flood, portable or temporary structures can damage other structures by direct impact, or lodge in the stream channel in such a way as to form a temporary dam, raising flood levels behind them. Siting housing areas of brush or forest land subject to wildfire can increase the damage potential to life and property. The same is essentially true of any structures placed near known fault lines in active earthquake zones, or in coastal or inland river areas subject to frequent wind and water damage from hurricanes.

Variables to be Measured. Each type of hazard has its own set of variables that influence frequency of occurrence and severity. For floods, changes in volume of the overland flow of water, changes in sediment deposits in stream channels, or alteration of the floodplain cause variation in flood height and resultant damage levels. Baseline data can sometimes be obtained from gauging stations that record the magnitude of the increased stream flow resulting from runoff associated with the storm. Changes in the infiltration rate of water cause changes in the volume of surface runoff from overland flow and in the amount of sediment carried into the stream. The resulting change in return frequencies of certain levels of flooding is the critical determinant of impact.

Earthslide-prone areas are those characterized by unstable slopes and land surfaces which—because of a history of actual occurrences, geology, bedrock structure, soil, and climate—present a significant hazard potential. The variables here are the extent to which soil and rock strata are exposed to wetting, drying, heating, and cooling processes; the slope gradient of the cut, which exposes the relevant stratum; and changes in internal earth stresses caused by surface or subsurface loadings, e.g., blasting, heavy machinery operation, and installation of footings and foundations.

The variables associated with wildfire are changes in flammability of the organic matter on the ground (duff) and the areal extent of the activity. Changes in wind velocity near the ground, depth of the duff, and moisture content of the duff influence its capacity to support combustion and the intensity with which it will burn once ignited. The size of the activity, in terms of changes in the volume or area of standing timber, in the number and value of physical facilities, and in the number of people housed or working in areas susceptible to wildfire, influences both the probable incidence of wildfire and the magnitude of the resultant damage.

How Variables are Measured. Few, if any, of the variables associated with baseline data on natural hazards are subject to measurement by the layman. It is even more difficult to project changes in the variables over time as a result of specific activities.

For floods, the assistance of an expert hydrologist is required to relate rainfall intensity (rate over time), infiltration capacity of the soil (the maximum rate at which soil in a given condition can absorb water), overland flow (rainfall excess that reaches stream channels as surface runoff) and its effect on channel depth, and the resulting increase in flow rate over time (hydrograph) which would yield a certain flood height and attendant damage level. The major variable—change in soil infiltration capacity—is influenced by such diverse and interrelated factors as interception of rain by trees and buildings, depth of surface detention of water and thickness of saturated soil layer, soil moisture content, compaction due to machines and animals, microstructure of the soil, vegetative cover at or near the surface, and temperature. No single formula can be used to relate changes in these variables to a specific change in the infiltration capacity of soil without the judgment of a hydrologist familiar with the watershed and area in question.[4] The nearest district office of the Corps of Engineers, Civil Works Division, or the U. S. Geological Survey should be able to assist in obtaining baseline data and in projecting the effects of various activities on flood heights and return frequencies.

The relative tendency of an area to have earthslides is not subject to simple measurement; the forces which cause an earthslide and the extent of their interactions are extremely complex. To the expert geologist, the type of geologic structure common to the area, the type of bedrock, soil structure, height of water table, type of surface material, degree of natural slope, and past history indicates whether an area is prone to earthslides. Such general information can often be obtained from the U. S. Geological Survey, from state geologists, or from local universities. If the area is prone to earthslides, an engineering analysis should be made to determine whether physical changes that result from the activity are likely to increase or decrease the probable incidence with which slides may occur. The services of both a soil and civil engineer would be required for a thorough analysis.

Baseline data on the conditions and occurrence of wildfire should be available directly from the nearest office of the U. S. Forest Service or from the state forester's office. These records usually include or can be correlated with other data relating to the thickness of the duff, the relative humidity, number of days since the last rain, wind velocity, and other local factors which, in combination, give the fire-danger rating. Local foresters specializing in the calculation of fire-danger ratings could assist in projecting the change the activity would have on the previously identified specific variables, i.e., wind velocity near the ground, depth of duff, and moisture content of the duff. Any change in the area (acres) susceptible to wildfire should also be measured. This can be done with before and after overlays of the area, prepared from maps, aerial photographs, or site plans. Through the use of a planimeter, the size of the area for each can be determined. The assistance of an engineer may be required to make this calculation.

Evaluation and Interpretation of Data. For flood hazards, the magnitude of the impact of a change in infiltration capacity of the soil and the attendant change in rate of surface runoff on flood stage height and of return frequency needs to be evaluated. A more sophisticated analysis could relate the change in flood height and return frequency to potential dollar losses or losses of human life, taking into account existing structures that might be affected, as well as any new ones to be located in the floodplain. This analysis could probably best be made by insurance underwriters associated with the National Flood Insurers Association or by the Federal Insurance Administration of the U. S. Department of Housing and Urban Development.

Evaluation of changes in potential incidence of earthslides is less straightforward. The areas where earthslides are most likely to occur should be evident from the previously recommended engineering analysis. The impact of a slide in a particular area could be calculated in terms of the dollar value of physical damage to structures, loss of life, and the ecological damage to watercourses and vegetation. A team of engineers, geologists, and insurance underwriters could develop risk factors associated with changes in the potential incidence of earthslides.

Just as with other natural hazards, wildfire has two aspects to be separately evaluated—the change in potential incidence and the amount of damage that might result from an occurrence. Again, the considerations are complex and not amenable to one-dimensional evaluation and interpretation. The change in incidence is related to change in flammability and areal extent of the duff, to greater or lesser numbers of people in the area, to the nature of the proposed activity, and to measures taken to prevent or reduce wildfire damage. A team of foresters and fire insurance underwriters should be able to develop risk factors associated with the change in potential incidence and intensity of wildfire and then estimate property damage or the loss of life that might result, both with and without the project.

Special Conditions. If increases in flood heights and frequencies are likely to adversely affect floodplains where extensive industrial, commercial, or residential development already exists; if increased incidence of earthslides is likely to damage population areas and/or cause severe ecological damage; or if residential or prime-timber producing areas are subjected to higher risks of damage from wildfire—particularly if any of the effects are felt outside the confines of the activity—controversy over the projected magnitude of the impacts is almost certain to develop. In such instances, an interdisciplinary team of qualified professionals is needed to develop and substantiate these projections.

Geographical and Temporal Limitations. Geographic limitations on natural hazards have to do with observed frequencies of occurrence, e.g., hurricanes are most likely to affect Gulf and Atlantic coastal areas; earth-slides are unlikely to occur in areas of relatively flat terrain; and earthquakes occur more frequently and with greater severity along known geological fault lines. While floods and wildfire can occur almost any-where, the frequency and severity of lightning storms in mountain regions of the western states increase the incidence of wildfire in that geographical area. There are some general temporal limitations for natural hazards: wildfire is most likely to occur in the summer and fall when the moisture content of living vegetation and the duff is lowest; floods of greatest severity occur with a certain predictability in the spring, but flash floods can take place at almost any time of the year; the hurricane season is considered to be summer and early fall; and earthslides of various types most often occur in the winter and spring. Temporal limitations do not seem to apply to earthquakes.

Mitigation of Impact. Primary mitigation techniques for hurricanes and earthquakes center around the avoidance of areas where these hazards occur with sufficient frequency and intensity to cause severe damage and the use of proofing techniques in the construction of physical facilities. Proofing techniques include the use of "floating" foundations and height restrictions in earthquake zones and increased foundation height, wall strength, and roof support in areas periodically subject to hurricanes.

The frequency and/or severity of flooding can be held to a minimum by prohibiting any construction activity or land use that restricts the flow of water in natural channels or that reduces the floodplain area that overflow waters during times of flooding. Generally speaking, all forms of temporary structures should be banned from the floodplain, and all permanent structures should be raised to a height above the level which flood waters can be expected to reach once every 100 years (100-year flood). No temporary dwelling units—mobile homes and the like—should be permitted in the floodplain.[5] Increases in surface runoff can be miti-gated by disturbing the existing vegetation and natural contour of the land as little as possible. Installation of underground drainage structures helps to reduce sediment loads (overland flow is reduced) but not total runoff volume.

Earthslides can be mitigated by avoiding areas with a high probability of incidence or those where proposed activity will significantly increase their probability. Engineering plans can be drawn to reduce the area of exposed strata subject to earthslides, reduce the inclination of slope of earth cuts on fills below what might otherwise be acceptable, provide physical support for exposed soil or rock faces, concentrate or distribute—as appropriate—the weight loadings of foundations to areas or strata better able to support that weight, use small charges for blasting, and restrict the movement of heavy machinery during the construction phase.

The effects of wildfire can be mitigated by clearing fire lanes in strategic locations and building restricted-access roads into areas having a high probability of wildfire incidence. Removal of live vegetative cover, which permits the drying forces of wind and sun to interact more directly with the duff, should, if possible, be avoided. In timber-harvest operations, the removal from the woods of as much of the total tree as is commercially possible to use will reduce the amount of vegetative logging debris left to contribute to depth and flammability of the duff. Restrictions on the use of areas during periods of high fire danger is another type of mitigation technique. Also, buildings should be sited (on the prevailing downwind slope) and roads constructed (more than one access and egress point) so as to minimize physical damage and loss of life if a wildfire should occur.

Secondary Effects. Activities that increase the risk of occurrence of natural hazards also have secondary impacts on various social and economic factors. General feelings of security and well being may be reduced by the increased threat of potential disaster. These psychological effects would be experienced most severely by individuals whose lives and property would be affected, should the disaster occur. Economic effects also could result in the forms of increased insurance premiums or changes in property values as hazard risks increase.

Other Comments. The impact of human activities in areas subject to hurricanes and earthquakes has not been treated in detail. The most appropriate measure of impact in such cases is the change in the number of people and in the dollar value of physical facilities exposed to these hazards as a result of the activity.

Land Use Patterns

Definition of the Attribute. Land use patterns are natural or imposed configurations resulting from spatial arrangement of the different uses of land at a particular time. Land use patterns evolve as a result of (1) changing economic considerations inherent in the concept of highest and best use of land, (2) imposing legal restrictions (zoning) on the uses of land, and (3) changing (zoning variances) existing legal restrictions.

The critical consideration is the extent to which any changes in land use patterns resulting from an action are compatible with existing adjacent uses and are in conformity with approved or proposed land use plans. The most recent guidelines on the content of environmental impact statements indicate that "where a conflict or inconsistency exists (between a proposed action and the objectives and specific terms of an approved or proposed Federal, state, or local land use plan, policy, or control), the statement should describe the extent to which the agency has reconciled its proposed action with the plan, policy, or control, and the reasons why the agency has decided to proceed notwithstanding the absence of full reconciliation".[6]

Activities that Affect the Attribute. Changes involving transportation systems (roads, highways, airports, etc.), water resources projects, industrial expansion, and changes in the working or resident populations are examples of activities likely to induce changes in the pattern of land use and create compatibility problems with adjacent uses. The building of new, or the expansion of existing, facilities through a program of land acquisition would be an activity likely to result in a conflict with approved or proposed federal, state, regional, or local land use plans. If such a conflict exists, it is quite possible that a compatibility problem with adjacent uses will also emerge. Recreational opportunities and second home or resort area development are other areas where land use conflicts are evolving.

Source of Effects. Activities involving land acquisition will conform or conflict with approved or proposed federal, state, regional, and local land use plans in relation to whether such plans exist at all, their detail, and the specific use of the acquired land. For example, if an agency purchased land for the construction of an office building in an area specifically designated for residential use by an approved zoning ordinance, there would be a direct conflict with a land use plan. Conversely, if the land were purchased as a site for the construction of family housing units, there would be no apparent conflict. These considerations were reflected in the decision rendered by the U. S. Court of Appeals of the District of Columbia in the case of Maryland Planning Commission v. Postal Service, August 1973.

In terms of changes in land use compatibility patterns, increased or decreased noise levels could have a decided impact. If an industrial-type activity is established at a location that was previously administrative in nature, the attendant increase in rail and truck traffic, particularly if routed near or through residential areas adjoining the site, could result in increased noise levels that might be incompatible with the existing use. Even greater noise problems affecting land use compatibility patterns arise in activities involving airfield construction or expansion, or modification of flight patterns.

Military installation closings resulting in the working and resident populations being reduced almost to zero would usually have a decided impact on the land use patterns of nearby private property. These changes might not be easily perceived at first. Residential and commerical areas would remain, but their intensity of use would probably be sharply curtailed. Portions of such areas might eventually revert to a lower use; the structures possibly razed and the land permitted to return to open space or some non-intensive form of agriculture. The issue of compatibility with adjacent uses might arise if the use revision took place in a random and essentially uncontrolled fashion.

Large increases in a project related labor force at a given location would almost certainly have repercussions on land use patterns in the area. An example would be the introduction into nearby areas of residential structures that are basically unsuited for such development. Mobile home parks or high-density apartment complexes might be sited adjacent to the approach pattern of aircraft runways on what was previously agricultural land. This could come about if variance to zoning ordinances was granted by some local governments in an attempt to encourage population growth in their political jurisdictions.

Activities which influence changes in land use patterns certainly do not always do so adversely. There may be compatibility conflicts in the existing land use pattern which would be ameliorated by other activities. An influx of people (with an appreciation of planning) into an area having no comprehensive zoning ordinances or land use plans could result in the formulation and adoption of such ordinances or plans. Over time, this could result in more compatible land uses in the area surrounding the activity.

Variables to be Measured. Compatibility of use between one parcel of land and adjacent properties involves variables such as type and intensity of use (residential, commercial, industrial, transportation, agricultural, mineral extraction, and recreational, and sub-breakdowns within each that reflects use intensity), population density, noise, transportation patterns, prevailing wind direction, buffer zones, and aesthetics. For example, a high level of residential/transportation land use compatibility would be evident where a single family home is set back 30 feet from a two-lane street having a traffic volume of 20 cars per hour which travel at an average speed of 25 miles per hour. Conversely, considerable incompatibility would exist if the same house is set back the same distance—with no intervening barriers—from a four-lane highway with a traffic volume of 2000 vehicles per hour, the majority of which travel at 55 or more miles per hour.

Conformity of a proposed new use of land with approved or existing land use plans is determined by whether a plan exists for the area in question, and if so, whether the proposed use conforms with the ones permitted in the plan. This is a very straightforward relationship unless attempts are made to correlate use/plan conflicts with variances under which precedents for change may have been set.

How Variables are Measured. Because the constraints that influence compatibility vary widely with the types of land use involved and the spatial arrangement of one with another, variables (such as traffic flow, population density, noise levels, depth/width/area of buffer zones, and constituents and quantity thereof in air/water/solid effluents) are subject to physical measurments by engineering and planning professionals. Even aesthetic qualities are subject to a somewhat objective measurement by landscape architects. With respect to compatibility of use, however, measurement alone does not indicate the magnitude of the impact. It is the relationship of these variables to one another in the context of their specific spatial arrangement that determines compatibility.

Measurement of variables reflecting conformity with a land use plan is essentially a yes-no proposition. A plan with which the proposed use can be compared either exists or does not. If a plan exists, the proposed use either conforms or conflicts with its provisions. In practice, the assistance of a spatial planner/zoning expert would probably be required if the proposed use is complex or if the plan is couched in legal terminology. Land use plans may be prepared at all levels of local governement: incorporated towns and municipalities, townships, and counties; by regional planning agencies (for agencies in specific areas, refer to *Regional Councils Directory*, published periodically by the National Association of Regional Councils); by state departments of planning, development, and natural resources (for specific state-by-state information, refer to *A Summary of State Land Use Controls*, published by Land Use Planning Reports, September 1973); and by federal land management agencies, such as the Bureau of Land Management, the National Park Service, the Bureau of Indian Affairs, the Bureau of Sport Fisheries and Wildlife, the Bureau of Reclamation, the Corps of Engineers (Civil Works Division), the Tennessee Valley Authority, and the Atomic Energy Commission.

Evaluation and Interpretation of Data. Discussion of the variables involved in land use compatibility attempts to convey the idea that there is no simple way to relate these variables and arrive at a compatibility index. While planning standards exist, the way they are applied in practice varies considerably from one political entity to another, from one geographic area to another, and with the types of existing and proposed uses. The assistance of a city and regional planner with a background in the spatial arrangement of land uses would be essential in measuring and analyzing interactions among variables and, subsequently, in interpreting the results in terms of the relative compatibility of the uses.

For reasons of continuity, evaluation and interpretation of whether a proposed use of certain parcels of land conforms or conflicts with existing or proposed land use plans was included in the previous discussion on the measurement of variables.

Geographical and Temporal Limitations. There appear to be no geographic limitations directly influencing the compatibility of adjacent uses of land. On the other hand, geographic boundaries of political entities govern the areal extent of the particular land use plans which the activity may impact.

Temporal considerations relate to the problem of projecting how land use patterns are likely to evolve as a result of a proposed activity. The period of analysis usually used is the expected beneficial life span of the project.

Mitigation of Impact. Compatibility between adjacent land uses can best be assured by providing an open-space buffer zone between the proposed activity and nearby properties where any significant degree of incompatibility is likely to result. The width/depth/area of this buffer zone should not be excessive, since to make it so could be construed as an inefficient use of land. As for mitigating the impact of changes in existing uses among adjacent off-post parcels of land likely to evolve as a result of the proposed activity, officials of affected local political entities and regional, state, and federal agencies could be apprised at an appropriate time of the projected impacts. They would then have the opportunity to change existing, or enact new, land use plans.

Mitigation of conflicts between a proposed use of land and proposed existing land use plans can best be accomplished during the planning stage. Obviously, it would be most desirable from an environmental standpoint to locate the activity where no conflict in use would exist. If this is not feasible, discussions could be held with representatives responsible for the plans, with a view toward resolving the conflict through the granting of a zoning variance or plan modification. Even if no satisfactory agreement can be reached, the fact that such discussions were initiated and conducted in good faith might have a positive impact on any future controversy or litigation.

Secondary Effects. Just as direct impacts on many biophysical and socioeconomic attributes induce effects on land use, direct effects on land use result in secondary effects on other biophysical and socioeconomic attributes. Transportation projects, for example, may concentrate air or ground traffic with resultant increases in levels of air pollution and noise production. Population shifts result in changes in demand for utilities (water supply, sewage treatment, electricity, etc.), and affect wholesale and retail markets and community services (police, fire, schools, etc.). In essence, land use designation can be related to all areas—air, water, land, ecology, sound, human, economic, and resources.

Other Comments. On the surface, it would appear that proposed land use plans, policies, or controls, as well as those which generally address land use without supportive legal instruments (ordinances, laws, administrative rules) would not be as binding—or taken into account to the same degree—as would those specifically and carefully drawn, officially enacted or promulgated, and having the support of legal precedent. However, the language of the previously quoted CEQ guidelines is rather unequivocable. For an impact assessment/statement, no differentiation is made between approved and proposed plans, policies, and controls. Time and precedent will determine how this element of the guidelines is interpreted in practice.

REFERENCES

1. *Design of a System for Evaluating the Environmental Impacts of Highways in Georgia* Battelle Columbus Laboratories, Report to the Georgia Department of Transportation, February, 1973.

2. *Rainfall Erosion Losses for Cropland*, Agricultural Research Service, *USDA Agricultural Handbook No. 282.*

3. *Control of Erosion and Sediment Deposition from Construction of Highways and Land Development*, U.S. Environmental Protection Agency, 1971.

4. Brater, E. P., and C. O. Wisler, *Hydrology*, Second Edition, John Wiley & Sons, New York, 1959.

5. *Code of Federal Regulations*, Title 24, Chapter VII, Subchapter B, Part 1910.

6. *Code of Federal Regulations, Federal Register*, **38**, *Number 147*, Part II, Title 40, Chapter V, Section 1500.8 (August 1, 1973).

ECOLOGY

The characteristics of man's environment are intimately related to the nonhuman ecology that surrounds him. Problems that affect lower level elements in the ecological system may ultimately affect man. For example, the accumulation of pesticides and heavy metals in lower levels of the ecological system may be harbingers of dangerous levels of these materials in man.

In addition, despite progress that man has made in providing for his needs, the total ecological balance of the environment is crucial to the viability of man. For this reason, species diversity and balance must be maintained. Convincing evidence exists that species diversity in an ecosystem is closely related to the stability of that system, with increasing species diversity indicating an increased ability of the ecosystem to resist disturbance and stress. Evaluation of impacts on a given ecological system should include an assessment of the effect of proposed alterations of the environment on species diversity, based on existing information or on special field studies.

The attributes that have been identified to describe the "ecology" resource are:

Large animals
Predatory birds
Small game
Fish, shellfish, and waterfowl
Field crops
Threatened species
Natural land vegetation
Aquatic plants.

Large Animals (Wild and Domestic)

Definition of the Attribute. Large animals are those, both wild and domestic, that weigh more than 30 pounds when fully grown. Common wild animals falling into this category are deer, bear, elk, and moose. Domesticated animals of this size include horses, sheep, cattle, swine, and goats.

Activities that Affect the Attribute. Since most large animals (except for some which are quite rare, i.e., cougars, wolves, etc.) are browsers or grazers, activities having the greatest effect upon them are those which diminish the animals' vegetative food supply or otherwise make inhospitable to them all or portions of the area over which they range. Examples of such activities are construction of new facilities (roads, fences, buildings, etc.), field training, and encroachment into wildlife habitat by vehicular traffic or recreation activities.

Source of Effects. Vegetative and other forms of cover—for traveling, eating, and watering, sleeping, breeding, and rearing of young—are required by all wild animals if they are to thrive in an area. Construction activities which result in the clearing of underbrush by burning or other physical means can reduce the available range over which large animals forage. Likewise, application of herbicides can reduce both cover and food, unless utilized in programs specifically designed to increase cover and food. Acquisition of new land for various activities, if such land was previously used for the grazing of domestic livestock, can reduce the total area available for that purpose in a particular locality. Noise can cause large wild animals to leave or avoid a particular area. Fencing can restrict the movement of animals, either denying them access to food and water areas or keeping them penned within an area smaller than that required for their well-being.

Variables to be Measured. The most direct variable is animal population. The type (species) and number of large animals should be determined. To arrive at the magnitude of the impact on the population, the change in the amount (acres) of land suitable for large animal habitat must be determined. A relative measure of the increased noise generated by man's extensive intrusion into wild, remote areas where he formerly ventured only as a hunter or herdsman should be made. Intense and prolonged noise-generating activity can sufficiently change the habits of large animals to cause them to vacate an area, at least temporarily, until man's activities are reduced or the animals become accustomed to them. Adjacent areas can be stressed by having to temporarily support greater populations.

How Variables are Measured. A census of large animal populations can be made by direct observation. If small, the entire area can be censused. If large, counts can be taken on random plots and projected over the total area of suitable habitat. Good observational and outdoor skills are required for many direct counts. In some areas of fairly open terrain, skilled photo-interpreters can take the census of large animals from aerial photographs. If direct observation is not practical or possible (lack of skilled people, large area, nature of the habitat or animal species), a local wildlife biologist affiliated with a federal or state wildlife agency should be consulted for his estimate of the population (numbers of domestic animals should be available from ranchers using the land). Wildlife specialists are professionally qualified to judge how noise and other nondestructive activities of man and vehicles affect the use of an area by large animals.

The change in acreage of a particular habitat type can be obtained from before and after overlays prepared from aerial photographic prints, project plans, or maps. Through the use of a planimeter, the size of these areas can be determined with the assistance of an engineer or surveyor. While a direct proportion can be made between the large animal population and acres of available habitat, it would be helpful to have a wildlife biologist review the calculations and determine the relative effect of the seasonal variations, etc. The following equation reflects the previously mentioned population/habitat area relationship:

$$\text{Future population} = \frac{\text{Present population} \times \text{future habitat acreage}}{\text{Present habitat acreage}}$$

Evaluation and Interpretation of Data. The increase or decrease of the large domestic animal population of an area can be interpreted on the basis of the resulting change in annual income. A more subjective evaluation must be made for wild animals. The number lost or gained relative to the number originally in the area is the most critical element. If any of these wild animals prey on smaller animals, the effect of the increase or decrease in that population should be considered. Not to be overlooked are the aesthetic value of large wild animals and the economic dividends which accrue to an entire region if the animals are subject to hunting. Neither of these two values can be readily quantified, and any judgment of their significance must remain highly subjective.

Special Conditions. If there is a long tradition of grazing rights for domestic livestock and these rights are to be withdrawn, the impact of the activity could become controversial—particularly if these rights had previously been exercised by native American tribes. If any of the wild animals are considered to be threatened (formerly categorized as rare or endangered)—regionally, nationally, or internationally—a reduction in their numbers as a result of some activity, particularly habitat alteration, would likely result in controversy. (The attribute write-up covering threatened species goes into greater detail on this subject.)

Geographical and Temporal Limitations. Concern about domestic animals and associated grazing rights is of significance primarily in the Western United States, where rights to use federal lands for this purpose still exist. As already noted, the impact of a particular activity on wild animals may be short-term, occurring only during the construction or direct activity period, when men and equipment intrude most heavily on the animal's home range.   Also, in alpine and high-plains areas, large animals have both a summer and winter range: a factor in determining their presence in, or absence from, an area.   The impact of the reduction in summer range would likely not be as severe, for example, as would be a reduction in winter range.

Mitigation of Impact.   The impact of activities on large animals can best be mitigated by intruding as little as possible on their habitat.   If such animals use the area where the activity will take place, the activity should be concentrated to the maximum extent possible in those parts of the area which they least often frequent.   During the planning phase of an activity, an attempt should be made to avoid extending into the home range of large wild animals.   If this is not feasible, the activity should be completed as quickly as possible, and regular and sustained use of the area over time should be minimized.   If land acquisition is necessary and a choice is possible, a productive range used by large domestic and/or wild animals should be avoided.

Secondary Effects.   Economic interests resulting from hunting-related business, and aesthetic qualities supported by the presence of wild animal species may be affected as a result of impacts on large animals.   Other secondary impacts may occur if natural predator-prey balances are upset by the activity.

Other Comments.   If the activity imrpinges upon the range of large wild animals that have previously been hunted in the particular area and if the activity will result either in closing that area to hunting or a reduction in the number of such animals available for annual harvest, sportsmen's clubs are likely to oppose the activity.

Predatory Birds

Definition of the Attribute.   Birds of prey are flesh eaters and obtain their food primarily by hunting, killing, and eating small animals, other birds, and fish.   Common birds in this group (Orders Falconiformes and Strigiformes) are hawks, owls, and vultures.   Less common are eagles, ospreys, and some of the falcons.   The California condor is quite rare.

Activities that Affect the Attribute.   Since birds of prey nest primarily in trees—sometimes in areas remote from human habitation—cutting of mature timber stands or the selected removal of overmature or non-commercial individual trees could result in a reduction in their numbers. Burning of brush or grasslands, applications of herbicides and pesticides, and the use of poisoned bait in animal-control programs are other activities that could directly affect the survivability of predatory birds. Activities resulting in intrusions by persons into or near nesting areas could affect these birds, particularly eagles, ospreys, condors, and some types of falcons that are less tolerant of man.

Source of Effects. The removal of nesting trees as a part of any general land-clearing program preceding construction activities or the selected removal of such trees in a forest-management "sanitation" cutting could destroy unhatched eggs or cause the death of birds too young to survive outside the nest. If suitable nesting habitat is not available elsewhere in the vicinity, adult birds may disappear from an entire area. Burning of brush and grasslands destroys the habitat and large numbers of the prey species (small animals) on which predatory birds depend for food. Similarly, application of defoliants could reduce the food and cover available for small animals and birds with a consequent reduction in their numbers. The effects of pesticides on birds of prey are not subject to universal agreement, but considerable evidence supports the proposition that the reproductive capacity of these birds may be reduced if sufficient quantities of DDT and other chlorinated hydrocarbon insecticides are concentrated in the food. As an example, egg shells can be weakened to the point where they break before the young are ready to emerge. Direct killing of predatory birds can result from their eating of poisoned bait (portions of animal carcasses) intended for coyotes, cougars, and other flesh-eating animal predators. Extensive outdoor activities resulting in the visibility of man and the noise of vehicular equipment and weapons firing, if conducted intensively over an extended period of time, or at frequent intervals, could cause birds to desert their nests. If the activity is sustained over a long enough period of time, adult birds may leave the area permanently.

Variables to be Measured. The number and types of birds of prey that nest and/or capture their food within the affected area should be determined. The change in the amount of available habitat (nesting and/or feeding) must be ascertained to estimate the numbers of birds which the existing habitat will support once the activity is completed or the project becomes operational.

How Variables are Measured. While a direct census of common birds of prey is possible in areas of limited size, the observational and general outdoor skills and the time required make it most impractical to conduct one. A usable population figure could be best obtained from a local wildlife biologist affiliated with a federal or state wildlife agency. Such a biologist of the Audubon Society, Isaak Walton League, or similar private wildlife conservation organizations, should be able to provide accurate counts of the less common species and the locations of their nesting and feeding areas.

The change in acreage of nesting and feeding habitats can be obtained from before and after overlays prepared from aerial photographic prints, mosaics, or topographic maps. Through the use of a planimeter, the size of these areas can be determined with the assistance of an engineer or surveyor. For the more common species of hawks and owls, the nesting and feeding habitats can be combined, and a direct proportion established, between the bird population and the number of acres of available habitat. The following equation reflects this relationship:

$$\text{Future population} = \frac{\text{present population} \times \text{future habitat acreage}}{\text{present habitat acreage}}$$

The relationship between available habitat and the generally larger, less numerous predatory birds is less direct and more subjective. If such species as the bald eagle, golden eagle, osprey, peregrin falcon, or California condor are present in an area, it would be best to solicit the opinion of expert wildlife biologists in determining what portion of the existing population would remain after the activity was completed.

Evaluation and Interpretation of Data.   The change in numbers of common birds of prey, as related to a particular activity and location, is an overall indicator of the change in habitat quality for other birds and animals within the area. Any substantial reduction in the numbers and types of hawks and owls would be generally indicative of a rather adverse ecological impact. As with the large wild animal attribute, any reduction of the less common species of avian predators could be expected to bring forth objections from private conservation organizations, as well as from federal and state agencies charged with their management and protection.

Special Conditions.   If any of the predatory birds of the area are considered to be threatened (formerly categorized as rare or endangered)—regionally, nationally, or internationally—any reduction in their numbers resulting from the activity, particularly habitat alteration, would likely be controversial. (The attribute write-up covering threatened species goes into greater detail on this subject.)

Geographical and Temporal Limitations.   Most of the large, less common birds of prey have very restricted geographic ranges. Maps showing these ranges are contained in most field guides to bird identification. A review of such range maps would reveal whether or not these species are likely to be found in the activity area. Special attention should be given to any short-term activities that might disturb the birds during their nesting season.

Mitigation of Impact.   The potential detrimental impact of human activities on the avian predator population can best be mitigated by locating the activity at places not considered a part of the habitat essential for the survival of these birds. This is best accomplished during the site selection planning stage of a project, rather than after a specific site has been chosen. Unless operational considerations are absolutely overriding, the habitat of the large, uncommon species should not be disturbed at all. Regular or sustained intrusions of men or equipment into nesting areas should be avoided to the maximum possible extent, especially while eggs are being incubated by the adults and until the young have left the nest. No known nests should be destroyed by the sanitation cuttings of noncommercial individual trees.

Secondary Effects.   Secondary impacts from an increase or decrease in predatory birds may be observed in the populations of animals upon which these birds prey. These animals may, in turn, have economic benefit through hunting-related business, or they may play significant roles in other ecologic relationships.

Other Comments.   If the existing habitat of the bald eagle, golden eagle, osprey, peregrin falcon, or California condor is threatened by an activity, the resultant controversy is likely to be intense, prolonged, and acrimonious.

Small Game

Definition of the Attribute.   Small game includes both upland game birds and animals which, as adults, weigh less than 30 pounds and are commonly hunted for sport. Some small game species falling into this category are rabbit, squirrel, raccoon, quail, grouse, and pheasant.

Activities that Affect the Attribute.   Since most small game animals and upland birds are very tolerant of humans, the activities most damaging to them are those which physically destroy their habitat (area in which all welfare factors such as food, cover, water, and space required for their survival and propagation are present in sufficient quantity and diversity). Land-clearing activities for buildings, road construction, etc., are most often the ones that significantly and adversely affect small game. Conversely, such game can be expected to return to formerly built-up areas, now abandoned, and allowed to revert to native vegetation. Distribution of poisoned baits used in rodent and predator control and use of herbicidal defoliants can also reduce small game populations, as can the use of certain pesticides.

Source of Effects.   The removal of native vegetation from, or the rearrangement of topography and surface features by the grading of, an area, denies small game the kind of habitat they require.   Without the food and cover provided by vegetation and irregular surface features, populations of small game diminish rapidly.   Conversely, they will quickly return to abandoned areas given over to native vegetation.   If poisoned bait is used, only a few small game animals and birds are likely to be affected, except in winter, when food is scarce and populations are at their annual minimum.   Herbicidal defoliants temporarily destroy small game habitat, and repeated applications can cause the permanent abandonment of an area. Persistent chemicals are accumulated in body tissues through ingestion of residues with food and water.

Variables to be Measured.   The small game population of the area to be affected by the activity must be censused.   Once this is accomplished, the number of acres of existing habitat must be determined, as well as the amount by which it will increase or decrease over time as a result of the activity.   The relationship between these variables and the attributes is fairly straightforward; the carrying capacity (wildlife population an area can support indefinitely without habitat degradation) is increased or decreased in direct population to the amount of available habitat.   While the quality of small game habitat existing before, and available after, the completion of an activity is an important variable, it is very difficult to quantify and will not be specifically discussed.   It will, however, enter into subjective evaluations and judgments.

How Variables are Measured.   While an accurate census of small game is difficult to make, usable estimates of the number of different species per acre of habitat can often be obtained from local wildlife biologists affiliated with federal or state wildlife agencies.

The change in acreage of small game habitat can be obtained from before and after habitat overlays prepared from aerial photographic prints, mosaics, or topographic maps. Through the use of a planimeter, the size of these areas can be determined with the assistance of an engineer or surveyor. A direct proportion can then be established between the small game population and the number of acres of suitable habitat. The following equation reflects that relationship:

$$\text{Future population} = \frac{\text{present population} \times \text{future habitat acreage}}{\text{present habitat acreage}}$$

Evaluation and Interpretation of Data.   The relative importance of a change in the small game population of an area is a very subjective judgment. If habitat is to be destroyed, significance should be attached to the relative amount and quality available in adjacent areas, as well as to the relative amount and quality of total habitat under control that will remain after the activity is completed.

Special Conditions.   If the activity will cause a significant reduction in the available small game habitat in an area subject to heavy hunting, the impact will likely be controversial to both sportsmen and, to a lesser extent, economic interests in the area. This could happen only if a prime small game hunting area is fenced and placed off limits to the general public.

Geographical and Temporal Limitations.   While it is unlikely that any small game species would fall into the threatened (formerly categorized as rare or endangered) category nationally, certain ones, such as grouse, woodcock, and turkey might be rare in some states or local areas. Many outdoor activities during nesting season often destroy eggs, the result of which may be a significant reduction of the game-bird population for one or more years.

Mitigation of Impact.   Activities affecting small game can best be mitigated by disturbing the vegetative cover and altering the physical contour of the land as little as possible. Selecting areas of poorer habitat quality and preserving prime areas will reduce the severity of the activities' impact on the small game population. Opening of large areas to the general public during certain periods of the small game hunting season (opening day and holidays, when hunting pressure is particularly heavy) will also tend to ameliorate the loss of areas formerly open to public hunting.

Secondary Effects.   Economic interests resulting from hunting-related business, and aesthetic qualities supported by the presence of wildlife may be affected as a result of impacts on small games. Other secondary impacts may occur if natural ecological predator-prey balances are upset by the activity.

Fish, Shellfish, and Waterfowl

Definition of the Attribute.   Fish are cold-blooded, aquatic animals that obtain oxygen through a gill system. They inhabit saltwater and freshwater bodies and streams and vary widely in size. Common species are minnows, sunfish, trout, bass, pike, salmon, tuna, and sharks.

Shellfish are aquatic animals that have an exoskeletal shell, rather than an internal vertebrate structure of backbone and ribs. Common freshwater and saltwater species are mussels, crayfish, clams, oysters, shrimp, crabs, and lobsters.

Waterfowl are birds which frequent and often swim in water, nest and raise their young near water, and derive at least part of their food from aquatic plants, animals, and insects. Ducks and geese are the most familiar waterfowl. Because of similar habitat requirements, the generally protected swans, herons, cranes, pelicans and gulls are also included here. The whooping crane is a frequently cited example of a threatened species that falls into the waterfowl category.

Activities that Affect the Attribute.   Since fish, shellfish, and waterfowl depend directly upon water for all or some facets of their existence, activities which affect water quality and water level have the greatest impact upon their well-being. Examples of particularly damaging activities are dredging, stream channelization, construction that exposes mineral soil and subsoil which is subject to erosion, disposal of untreated or insufficiently treated sewage in water courses, permitting toxic materials to drain into water courses without collection and treatment, disposing of cooling water in the ocean or in streams and lakes, application of pesticides—the residue of which may drain into water courses, draining of swamps or potholes, building of water-level control structures such as dams or dikes, and disposal or containerized toxic gases and residues at sea.

Source of Effects.   Dredging can temporarily displace the bottom organisms on which these categories of wildlife feed and can destroy spawning grounds. Stream channelization results in the removal of native vegetation which supports the insects eaten by fish. In addition, alteration of flow and substrate characteristics resulting from stream channelization can be as harmful as loss of vegetation. Certain species of fish are affected by even small amounts of solid material suspended in the water, a condition resulting from dredging or soil erosion. Some species of fish and shellfish are affected by siltation, which can both cut off their oxygen supply and reduce the availability of food. Discharge of insufficiently treated sewage may introduce disease-causing bacteria and viruses and reduce the oxygen content of the water—the life-support system upon which fish and shellfish are totally dependent. Insufficiently treated sewage also introduces nutrients which accelerate plant growth into the water, often affecting the quantity of available fish habitat by further reducing the oxygen supply. Toxic materials, such as oils spilled or draining into water courses, cause the feathers of waterfowl to no longer shed water, bringing about death from exposure. Toxic materials, such as mercury, can eventually be so concentrated in the food chain that fish are no longer safe for man to eat. Other toxic materials can cause the outright death of fish by damaging their gills and preventing them from extracting oxygen from water. The acidity level of water, if too high (pH 5 or less), or too low (pH 11 or greater), can cause similar gill damage. Increases in water temperature often cause sport fish to abandon the area to less desirable species of so-called rough fish such as carp. Rapid fluctuations in water temperature can kill fish outright. Pesticide residues draining into water courses and concentrating through the food chain may eventually become present in sufficient quantities in fish to cause their reproductive capacity and the survivability of young to be impaired. Pesticides can become even more concentrated in the tissues of fish-eating birds and animals. Draining of swamps or potholes is very detrimental to waterfowl, as it is near these bodies of water that reproduction, nesting, and the rearing of young take place. The artificial raising and lowering of water levels is often beneficial to wildlife habitat, if done at times consistent with needs for food and nesting cover. However, since changing of water levels is most often a flood-control requirement, fish and waterfowl habitat can be drastically affected by changes not in consonance with their needs. Depending on the lethality of the material, the rupturing of containers of toxic substances disposed of at sea could cause the destruction of all aquatic life in both the immediate area and other areas where the substance is transported by ocean currents.

Variables to be Measured. The variables to be measured for fish are those identified in the attribute descriptions involved with surface water quality. Some of these variables are dissolved oxygen content, coliform bacteria levels, acidity levels (pH), heavy metal concentrations, and insecticide concentrations which are detrimental to fish life.

While many substances (petroleum products, hydrogen sulfide, copper, and other metals) can taint shellfish and make them unpalatable for reasons of odor, taste, or color; pathogenic bacteria and viruses which they take up from the surrounding water may render them unfit for human consumption. Measurements of coliform bacteria present in the water provide a standard for determining when oysters, clams, and muscles can be safely eaten.

The main variable to be measured for waterfowl is change in available habitat. Quantity of suitable nesting habitat—which equates to the length of shoreline—is a heavy determinant of waterfowl population on a year-by-year basis, but it is difficult to relate the two exactly. Winter habitat is also important, but more difficult to quantify and to relate to increases or decreases in waterfowl.

How Variables are Measured. As indicated, measurements of water quality variables are discussed under surface water quality. Some acceptable general standards for maintaining a healthy aquatic fish habitat are that dissolved oxygen content should not fall below 5 milligrams per liter and that pH level should be maintained in the 6 to 9 range. The Federal Environmental Protection Agency (EPA) has published a draft report which specifies limits for pollutants within various water use categories, several sections of which deal with aquatic life. Prior to formal publication of the report, these standards are available in summary form in the *Environment Reporter*.[1]

While the standards for fish also apply generally to shellfish, coliform bacteria count is the important variable to be measured. Criteria for water from which shellfish are harvested are contained in the U.S. Public Health Service Manual, *Sanitation of Shellfish Growing Areas*. General standards for coliform bacteria are that the median most probable number (MPN) must not exceed 70 per 100 milliliters.

If the length of existing shoreline or its character is altered so as to render it unsuitable for waterfowl nesting, the amount of change should be determined. Before and after overlays of suitable wildlife nesting habitat along shorelines should be prepared from aerial photographs, projects plans, or maps. Through the use of a map measurer, shoreline length can be determined. The amount of change between present and future habitat can then be calculated. Additionally, the change in number of individual bodies of water between the two overlays should be noted. Once these data are obtained, they are still not directly convertible to change in the number of pairs of nesting waterfowl the habitat can support. This is a subjective judgment which only an expert wildlife biologist, e.g., from the U. S. Fish and Wildlife Service, can make.

Evaluation and Interpretation of Data. Although it is difficult to relate alterations in water quality to changes in fish and shellfish populations, changes in fecundity, population counts, and growth rates are often sensitive indicators of such alterations. Therefore, an attempt should be made to assess population changes that might result from proposed alteration of the environment. If water quality is degraded or improved to the point where commercial fishing activities are affected, the change in annual revenues derived from this source can be determined. If the change in water quality affects species associated with sport fishing, the number of miles of streams affected would provide some measure of the significance of the impact. If a prime sport fishing area is involved, economic gains or losses to businesses deriving a part of their income from fishermen might be an important consideration. Estimates of the effects of such changes might be obtained from federal or state wildlife agencies.

Changes in quantity of nesting habitat would definitely affect the number of waterfowl available for annual harvest. However, the effect is felt more in areas where waterfowl are hunted than where they nest. An expert from the U. S. Fish and Wildlife Service could provide an insight into the extent of the resulting environmental (ecological and economic) impacts.

Special Conditions. If activities will cause a significant reduction in the length of streams or areas of coastal waters suitable for sport fishing or in the amount of waterfowl nesting habitat, the impact will likely be controversial to sportsmen. This could happen if even small stretches of trout streams were to be affected of if prime fishing waters were placed off limits to the general public. Commercial interests would most likely oppose any intrusion into prime fish and shellfish areas or any reduction in the annual catch/harvest. If any waterfowl that are considered to be threatened (formerly categorized as rare or endangered)—regionally, nationally, or internationally—use the activity area for nesting, migration stopover, or feeding, significant controversy would probably result. (The attribute writeup covering Threatened Species goes into greater detail on this subject.)

Geographical and Temporal Limitations. The only geographic limitations on fish and shellfish relate to particular types found in the activity area and whether coastal estuaries or open sea areas are involved. Temporal considerations are those involved with conducting the activity during spawning, migration, or harvest seasons.

The critical region of waterfowl nesting habitat is generally considered to be in states adjacent to the Canadian border and in Alaska. This would, however, not be true of non-migratory waterfowl associated with estuaries and seacoasts. Activity which would disturb waterfowl during the nesting season and while the young are being reared would be most damaging.

Mitigation of Impact. Impacts upon fish and shellfish populations can be mitigated by restricting the input of polluting substances into water courses, estuaries, and the open sea. This can best be accomplished by insuring that wastewater treatment facilities of suitable capacity and design are constructed so as to be in operation by the time it is anticipated that waste products from the proposed project will be generated. If soil erosion is a problem, construction activities should be scheduled at times of the year when intense rainfall is least likely to occur.

Impacts on waterfowl from an activity can best be mitigated by disturbing the land/water interface in the area as little as possible. Vegetation along water courses should not be cleared indiscriminately. Neither should potholes or swamps be drained unless absolutely necessary for successful completion of the activity. Additionally, when a part of the activity involves water level control, changes in such levels should be programmed—to the extent it is possible to do so—in a way that will only minimally disturb nesting and feeding habitat. These considerations for the natural environment will help to insure that waterfowl habitat available for nesting and feeding is not appreciably diminished in either quantity or quality.

Secondary Effects. Economic interests resulting from businesses related to hunting and commercial and sport fishing activities would be affected by impacts to fish, shellfish, and waterfowl. Other secondary impacts may occur if natural ecological relationships are upset by the action.

Other Comments.   Water quality and fish and shellfish habitat go hand in hand. Any substantial degradation of the former will have a decided impact on the fish and shellfish populations relative to both quality and number. All aquatic oxygen-using (aerobic) organisms will be affected to some degree by decreases in water quality. The effect of an activity on fish and shellfish is a general indicator of the impact on the entire water environment.

Field Crops

Definition of the Attribute.   Field crops are those commercially cultivated by man for the primary purpose of providing food and clothing for himself and food for domestic livestock. Common field crops include corn, wheat, cotton, soybeans, and truck produce (tomatoes, melons, and table vegetables).

Activities that Affect the Attribute.   Since almost all land suitable for field crops is in private ownership, acquisition of that land—for whatever specific purpose—would take it out of agricultural production. Acquisition of prime agricultural lands for nonagricultural purposes is likely to have the greatest impact on field crops. Reservoir construction and operation, along with various runoff control projects, may affect the flooding of large areas of field crops. Application of herbicides on land adjacent to an agricultural area planted in field crops would have a more localized impact.

Source of Effects.   Diversity, both man-made and natural, is an important and valuable characteristic of ecosystems. If the area previously given over to field crops is to be built upon (the most likely reason for acquiring the relatively flat land that field crops usually occupy) or is to be used extensively for nonagricultural activities, the vegetative diversity will be reduced. Wildlife could also be affected, as many game and nongame animals and birds obtain food and cover from field crops. If the acquired land is allowed, through successive vegetative stages, to revert to the natural climax type of area, the impact might be ecologically beneficial.

Reservoirs and impoundments may raise groundwater levels, flooding the root systems and severely damaging crops. Other flow diversions may decrease probability of flood damage.

Herbicides applied by aerial spraying might carry onto adjacent agricultural land, killing crops with which they come in contact. While the area might be relatively small, the resulting damage could be highly controversial.

Variables to be Measured. The main variables to be measured are the number of acres of land now given over to field crops which would be taken out of production, and the percentage of that land which would be permitted to revert to natural vegetation. Since field crops and vegetation are both ecologically important, an assumption can be made that if one type is not unduly created at the expense of the other, each is of equal significance. The measure of ecological impact would then be determined by the loss of productive vegetative cover.

How Variables are Measured. Specific acreages of field crop land to be taken out of production by a land acquisition program could be measured directly, but it would be easier to obtain figures from local offices of the Agricultural Stabilization and Conservation Service (ASCS) of the U. S. Department of Agriculture. Land previously used for crops, but permitted to revert to natural vegetation, should be depicted on an overlay prepared to scale from a project plan, a map, or an aerial photograph. Through the use of a planimeter, the size (acreage) of that area can be determined with the assistance of an engineer or surveyor. The percentage of crop land reverting to natural vegetation could be derived from the following equation:

Percentage of land reverting to natural vegetation =
$$\frac{\text{area reverting to natural vegetation}}{\text{total area of crop land}} \times 100.$$

Only a general estimate is possible when determining field crop acreages that might be damaged by herbicidal spraying. Some of the variables involved would be the kind of application system used, wind direction, and velocity, and state of crop development. However, if these variables are reduced to an assumption that 500 feet is the maximum distance into the field that the herbicide could produce crop damage, the other variable involved would be the linear measure of crop land directly adjoining the area where the herbicide is to be applied. Solution of the following equation would provide a quantitative measure of this effect:

$$\text{Acres affected} = \frac{500 \text{ feet} \times \text{the linear measure of crop land affected (feet)}}{43,560 \text{ square feet}}$$

This effect could then be translated into economic terms through the use of another equation:

Economic returns foregone ($) =
$$\frac{\text{acres affected} \times \text{average yield per acre (bushel, ton, etc.)}}{\times \text{ selling price per unit}}$$

The ASCS should be able to provide acreage yields and selling prices of various field crops which might be affected.

Evaluation and Interpretation of Data.   The magnitude of the impact of the change in land use that results from crop land acquisition is related to the percentage of that land which will continue to support natural vegetation. The greater the percentage of field crop land that is built upon or otherwise taken out of vegetation production, the greater the impact. For crops damaged by application of herbicides, a measure of impact could be made by comparing the dollar loss of the destroyed crops to the annual value of that crop in the country or area concerned. Again, the greater the percentage of dollar loss is of the total crop value, the greater the impact.

Special Conditions.   If the crop land is especially productive relative to other crop land in the general area, or if the crop grown upon that land is of very high value, the impact may be greater than what would otherwise be anticipated.

Geographical and Temporal Limitations.   Because of climate and soil or other requirements, some field crops—particularly truck crops (avocados are a good example)—can be grown only in a very limited geographic area. If the crop land to be acquired is in such an area, a significant reduction in the local output of that crop might result. On the other hand, there are vast areas in the Western United States where conditions are not suitable for the cultivation of field crops. Land acquisition activity in those areas would not affect this attribute.

Herbicidal damage to field crops is greatest when the plant is growing fastest (spring) before the vegetative product (corn ears, grain kernels, bean pods) has matured. However, this is generally the time when herbicides will be used, because they have the greatest suppressive effect on vegetation at which they are directed.

Mitigation of Impact.   The detrimental impact of acquiring productive field crop land can best be mitigated by locating the activity in an area where very little land is given over to field crop production or where the farming enterprise is of marginal economic value. Some additional mitigation in the form of trade-offs is possible if a large portion of the crop land is allowed to revert to natural vegetation. This would be possible in buffer areas acquired to shield private lands from specific activities.

Mitigation of the impact of herbicidal applications could take the form of cutting vegetation and applying the herbicide directly to the stumps in those areas where field crops are directly adjacent to the activity. Further mitigation of impact is possible if the stump application of herbicides is done at a time when the adjacent field crops are vegetatively dormant. If spraying is a preferred method of vegetative suppression, it should be done at times when wind velocity is low and wind direction is such that the possibility of the herbicide carrying into the field crop area is minimal.

Secondary Effects. Loss of field crop production could have significant effects on local or regional economic stability, particularly if the loss is of long-lasting or permanent duration. Other effects on land prices and farm product availability could result.

Other Comments. If significant economic loss will result from the acquisition of crop land and its removal from agricultural production, farmers' organizations would be likely to actively oppose the project.

Threatened Species

Definition of the Attribute. Threatened species (formerly categorized as rare or endangered) include all forms of plant and animal life whose rates of reproduction have declined to the point where their populations are so small they are in danger of disappearing. Threatened species are classified as such on a state and national basis. A species classified as threatened within a state may occur only in limited numbers at very few locations within that state but be common in other states. National threatened species are those found only in very small numbers or those near extinction in the United States. Lists of threatened animal species are published periodically by the Bureau of Sport Fisheries and Wildlife of the U.S. Department of Interior.[2] Examples of more commonly known threatened species are the timber wolf, grizzly bear, Southern bald eagle, and whooping crane. Less commonly known threatened species include the black footed ferret, key deer (Florida), Devil's Hole pup fish, Florida kite, and Delmarva fox squirrel. While animal species are the ones most often in the public eye, there are probably many species of plants that would also qualify as threatened. The coast redwood is already a candidate for popular interest in this category, and others seem certain to gain public attention.

Activities that Affect the Attribute. These activities are basically the same, depending on the animal or plant species involved, as those mentioned under the Large Animals, Predatory Birds, and Natural Land Vegetation attributes. Refer to those sections if a threatened species' habitat is located within the geographic area that a specific action will affect.

Source of Effects. The source of the effects of various activities on threatened species of animals and plants is essentially the same as those listed for Large Animals, Predatory Birds, and Natural Land Vegetation. Refer to those attributes if the habitat of a threatened species is located within an area where the effects of a particular action will be felt.

Variables to be Measured; How Variables are Measured. The variables to be measured and the method of doing so are highly dependent upon the particular species of plant or animal affected. While the information contained in similar parts of the attribute write-up for Large Animals, Predatory Birds, and National Land Vegetation could serve as a general guide in the case of threatened species, the assistance of an ecological team of wildlife biologists, zoologists, botanists, and plant physiologists in accumulating relevant data would be almost a necessity.

Evaluation and Interpretation of Data. This function can be adequately carried out only by a group of professional ecologists familiar with the myriad details associated with the threatened species itself, its place in the ecosystem, and the nature of the particular habitat which is to be impacted. Logically, this team of ecologists should be the same group responsible for collection of the basic data on which the evaluation is to be based. However, an additional critical review of their conclusions by an eminent ecologist might help to insure public acceptability of those findings.

Mitigation of Impact. The primary way to mitigate the impact of activities on threatened species is to avoid any disruption—physical or biological—of their habitats which might result in a decrease in their populations. While it would be less damaging to disturb the habitat of a species classified as threatened by a state than one that is threatened nationally or one that is rare rather than endangered, these trade-offs are usually not feasible. It is best to avoid disturbing the known habitat of any threatened species.

Secondary Effects. As indicated above (in Special Conditions), many secondary effects occur along with the impacts on threatened species. In addition to their human and aesthetic interest value, some of these species may be of significant importance to the dynamic aspect of the ecosystems in which they are found. Their use as indicators of overall environmental quality should not go unnoticed.

Other Comments. If any activity has the potential of adversely affecting the populations of any threatened species, naturalist and wildlife groups are almost certain to vigorously oppose it in public hearings and/or in court.

If any question exists as to the presence of a threatened species—either intermittently or year-round—in the area of a project, local wildlife biologists or botanists should be called upon to verify that presence and to give a preliminary assessment of the impact of the activity on the population of this species.

Natural Land Vegetation

Definition of the Attribute.   Natural land vegetation is that which uses soil (as opposed to water) as its growth medium and which is not the subject of extensive cultural practices by man. Included in this category are a number of diverse groups of plants, including trees, shrubs, grasses, herbs, ferns, and lichens.

Activities that Affect the Attribute.   Any activities that affect land surface will affect the vegetation that grows upon it. Timber-harvest operations, land clearing activities prior to construction, burning, application of herbicides, off-road vehicular traffic, and the application of artificial paving materials are some activities that can cause adverse impacts on natural vegetation.   Abandonment of facilities can result in natural vegetation becoming reestablished through a series of successional stages.

Source of Effects.   Timber-removal operations employing inappropriate forest management methods can reduce the possibility of reestablishing fully stocked stands of the same species.   Without the protection of the forest canopy, shrubs and other plants left after timber removal may weaken and become prime targets for disease and insects.   Land clearing activities can cause the outright destruction of natural vegetation, and resulting soil erosion can inhibit its reestablishment. Improper use of herbicides can result in the destruction of nontarget species of natural vegetation and can disrupt the overall stability of the ecosystem.   Mechanized field training destroys lower vegetative forms outright, and the resultant soil compaction and erosion—each in its own way—can inhibit their reestablishment. Paving can deny native vegetation to large areas for extended time periods. As previously indicated, a reduction in the magnitude of activity at a particular installation or its closing can encourage the reestablishment of native vegetation.

Variables to be Measured.   The variables to be measured are the number of acres of native vegetation existing before and after the activity, as well as any significant vegetative changes that may develop.   A reduction in an area given over to native vegetation can result in increased soil erosion, a decrease in soil fertility, and a decrease in quality and quantity of wildlife habitat.   It can also accelerate the invasion of weeds and other undesirable pest species.   Reintroduction of native vegetation can—over time—

have the opposite effect. Successional change in vegetative type is slow, however, and the least desirable plant types are the first to become reestablished on a site after a major clearing activity.

How Variables are Measured.   The change in acreage of natural vegetation can be obtained from before and after overlays of vegetative types. The before overlay can best be prepared from recent aerial photographs. A photointerpreter can assist in differentiating and plotting the major vegetative types.   In this way, the total area of vegetation cover can be ascertained along with the subareas in each of the major types. The after-activity overlay should be prepared at the same scale, using the project plan to outline areas of existing natural vegetation which will be affected. The remaining total acreage in native vegetation by major type should then be determined. These calculations of acreage can best be done through the use of a planimeter. The percentage of original native vegetation remaining—both total and by major type—is then derived.

Evaluation and Interpretation of Data.   The magnitude of the impact of the activity on natural vegetation can be determined from the percentages previously given.   However, the specific changes and types of vegetation which would result from the activity could be projected only by a botanist or forester intimately familiar with the local area. Even more difficult to interpret objectively are the aesthetic considerations involved.

Special Conditions.   Destruction of natural vegetation in particularly fragile ecosystems that exist under extremely adverse environmental conditions—such as tundra and desert, can have greater impact than in an area with a more moderate climate.

Geographical and Temporal Limitations.   The only geographic limitations of impacts on this attribute occur in those rare areas of desert and bare rock where no native vegetation exists. No temporal limitations are evident.

Mitigation of Impact.   The best way to mitigate the impact of activities on natural vegetation is to design the project so as to restrict the area affected. Examples of other mitigation possibilities are to restrict land clearing activities to the absolute minimum, apply ecologically sound management practices in timber harvest and timber stand improvement, confine vehicular activities to designated areas and restrict expanding them into new areas, apply vegetation suppression techniques of controlled burning and herbicide application only when other methods are not feasible, and use crushed stone rather than asphalt or concrete for surfacing parking areas.

Secondary Effects.   In addition to economic gains from timber harvesting, natural land vegetation provides habitat for wildlife species, recreational areas of hunting, camping, and other pursuits, and countless other resources of both aesthetic and material nature.

Other Comments.   If activities result in the destruction of unique areas of natural vegetation, opposition can be anticipated from local and national naturalist organizations.   These natural areas are usually well known locally and are often cataloged at the state level by departments of natural resources.   Any activity that would alter these unique and rare areas of natural vegetation should be avoided to the same extent as one involving the habitat of a threatened species of wildlife.

Aquatic Plants

Definition of the Attribute.   Aquatic plants are those whose growth medium is primarily water, though they may be rooted in soil. They include free-floating plants such as phytoplankton, all surface and submerged rooted plants, and swamp and marsh vegetation whose roots are periodically or permanently submerged in water. Aquatic plants are essential elements in the food web.

Activities that Affect the Attribute.   Activities which cause changes in water level or water quality parameters have the greatest impact on aquatic plants.   Examples of particularly damaging activities are dredging, stream channelization, construction that exposes mineral soil and subsoil subject to erosion, disposal of untreated or insufficiently treated sewage in water courses, disposal of cooling waters in oceans and in streams or lakes, draining swamps and marshes, and building of water-level-control structures such as dams or dikes.

Sources of Effects.   Dredging can temporarily—and sometimes for long periods—displace rooted and bottom-dwelling aquatic plants.   Stream channelization removes all stream-side vegetation.   Erosion can cause increased sediment loads sufficient to restrict the sunlight on which aquatic plants depend for photosynthesis.   The discharge of insufficiently treated sewage into the water courses induces excessive aquatic plant growth. Increases in temperature also tend to accelerate aquatic plant growth, particularly algae.   The draining of swamps and marshes reduces the area in which aquatic plants can survive.   Changes in water level can cause the destruction of aquatic plants, either by exposing their roots to the drying influence of sunlight or air or by flooding to levels which deny air to bank- or marsh-dwelling species for long periods of time.

Variables to be Measured. The essential variable is the change in amount of water area suitable for the growth of aquatic plants. There are two elements to this variable: changes in water surface areas and changes in those elements of water quality which accelerate or restrict plant growth. Any changes in the kind of vegetation and its productivity can influence all other organisms that depend upon it for food.

How Variables are Measured. The only direct measurement that can be readily made is the quantity of total aquatic plant habitat available before and after the activity. This can be done by an expert photointerpreter, who should prepare before and after overlays from large scale aerial photography. He can then measure the acreage in those areas with a planimeter. The percentage change in total available aquatic habitat is then derived.

The quality of the water habitat existing before and after the project can be ascertained only by intensively examining the aquatic plant life, measuring the various water quality parameters affecting plant growth, projecting changes in water quality that will result from the activity, and projecting the changes in aquatic plant habitat that will follow. This is a complex procedure which can best be accomplished with the assistance of an interdisciplinary team of biologists, botanists, zoologists, ecologists, and engineers.

Evaluation and Interpretation of Data. Since it is not possible to directly measure the change in the quality and quantity of aquatic plant life, only a very imprecise measure of impact can be obtained from the change in water area. Generally, if the percentage change in aquatic plant habitat exhibits a value greater than 20, an attempt should be made to measure the qualitative change as well. Further, if changes in the water quality parameters measured or projected under surface water attributes indicate increased nitrogen and phosphorous concentrations, increased water temperature, decreased water flow, or high sediment loads, the advice of ecological experts should be sought relative to the extent of the impact on aquatic life.

Special Conditions. If the change in quantity, quality, or type of aquatic vegetation will result in waters being rendered unfit for swimming or will cause a reduction in the game-fish or commercial-fish populations, greater controversy over the project is likely to result since the impact will be more directly felt by the general public.

Geographical and Temporal Limitations. The only geographical limitations on aquatic vegetation are the particular types native to certain areas. Temporal considerations do not appear to be significant.

Mitigation of Impact. Impacts on aquatic plant life can best be mitigated by minimizing the input of nutrients, erosion products, and heat into water bodies. This can be accomplished by assuring that wastewater treatment facilities of appropriate size are constructed so as to be in operation by the time increased amount of nutrients is scheduled to be generated. If soil erosion is a problem, catch basins can be constructed to permit the settling out of suspended solids prior to the runoff water reaching natural water bodies. (The attribute write-up covering Erosion goes into greater detail on this subject.) Additionally, construction activities can be scheduled at times of the year when intensive rainfall is least likely to occur. Cooling water can be processed or stored in artificial ponds until the difference in temperature between it and the receiving water is more nearly equal.

Swamps and marshes should not be drained unless such action is absolutely necessary for the successful completion of the activity. Artificial changes in water level should be minimized and programmed during the fall and winter, when the plants are dormant. If herbicides are used to suppress excessive aquatic plant growth, they should be applied selectively and in amounts that will reduce the undesirable species but not kill all aquatic plants.

Secondary Effects. Since aquatic plants are essential elements in the food web, adverse impacts to these elements will also be reflected in impacts to higher order consumers (fish, animals, and man). Excessive growth of aquatic plants, on the other hand, can choke waterways and recreational areas, with resultant induced reduction of economic, social, and aesthetic benefits.

Other Comments. Water quality and quantity are directly related to the suitability of water bodies for desirable aquatic plant growth. Introduction of pollutants will reduce plant productivity and plant species diversity and result in an aquatic plant community composed predominantly of pollution-tolerant forms. This will, in turn, have a decided impact on the fish population that inhabit the waters. Changes in the food web can have impacts throughout the ecosystem, but these are often not completely understood.

REFERENCES

1. *Environment Reporter* **4,** *No. 16:* 663 and 669, August 17, 1973.

2. "Threatened Wildlife of the United States," 1973 Edition, *Resource Publication 114*, U.S. Department of Interior Fish and Wildlife Service, Bureau of Sport Fisheries and Wildlife, U.S. Government Printing Office, Washington, D.C., March 1973.

SOUND

The level of sound (noise) is an important indicator of the quality of the environment.   Ramifications of various sound levels and types may be reflected in health (mental and physical) and/or in aesthetic appreciation of an area.   Because of the important consequences of a too-noisy environment, this resource, sound, is examined separately rather than under various other resource categories.

The sound (noise) in an environment is indicated by many attributes, but the selected ones are:

Physiological effects
Psychological effects
Communication effects
Performance effects
Social behavior effects

Physiological Effects

Definition of the Attribute.   Noise can affect the physiology of the human body in three important ways:

Internal bodily systems
Hearing threshold
Sleep pattern.

Internal bodily systems are defined as those physiological systems essential for life support, i.e., cardiovascular (heart, lungs, vessels), gastrointestinal (stomach, intestines), neural (nerves), musculoskeletal (muscles, bones), and endocrine (glands).   Noise stimulation of nerve fibers in the ear may indirectly harmfully affect these systems.   High-intensity noise (e.g., artillery fire, inside tracked vehicles) constricts the blood vessels, increases pulse and respiration rates, increases tension and fatigue, and can cause dizziness and loss of balance.[1-3]   However, these effects are generally temporary; and, to some extent, adaptation does occur.   The process of adaptation is in itself indicative of an alteration in body functions and is therefore undesirable.   Persistent effects have been reported (people working for years in high noise level environments), but these reports remain unproven at this time.[2] High noise levels can also reduce precision of coordinated movements, lengthen reaction time, and increase response time, all of which can result in human error. Those people (particularly the elderly) who have circulatory problems, chronic heart disease, or tension-related diseases may be adversely affected by high noise levels.[1]

Hearing threshold is defined as the lowest sound level or loudness of a noise that can be heard. The lower the sound level that can be heard, the lower the hearing threshold. If the sound level necessary for a noise to be heard (or the hearing threshold) is higher than normal, then hearing loss or partial deafness is indicated. Noise can cause temporary or permanent hearing loss (i.e., an increase in the hearing threshold) and can cause ringing in the ears (tinnitus). Hearing loss can be temporary, in that the ear recovers relatively soon after the termination of the noise. Hearing loss of any degree is serious because accidents can occur if warning signals, commands, etc. cannot be heard or understood. In addition, hearing loss is undesirable from social, economic, psychological, and physiological points of view.

Sleep pattern is defined as a natural, regularly recurring condition of rest, and is essential for normal body and mental maintenance and recuperation from illness. Noise can affect the depth, continuity, duration, and recuperative value of sleep. The disruption or lack of sleep results in irritability, often irrational behavior, and the desire for sleep. Even a shift in the depth of sleep can result in fatigue. Also, while suffering or recovering from illness, rest and sleep are essential to health and recovery. Thus, it is important for noise to be kept at a minimum during night hours.

Activities that Affect the Attribute. Most activities cause some level of noise, but the most serious impacts are:

- *Construction.* Construction projects create noise through the use of vehicles, construction equipment, and power tools. The noise affects the operators, personnel, and communities near the site, and the people near transportation routes to the site.

- *Operational activities.* The operation of most types of air/surface vehicles, machinery, power-generating equipment, and weapons will generate noise. Maintenance and repair produce noise through the use of all types of tools, and when a number of noise sources are operating at the same time in the same general area (e.g., a vehicle repair shop).

- *Military training.* Training courses and exercises which use any type of vehicle, weapon, power tools, appliances, and machinery create noise for operators, military and civilian personnel, and, in large-scale exercises, can affect nearby civilian communities.

• *Industrial plants.* The machinery and tools contained in these plants are a significant source of noise to the personnel and, if noise levels are sufficiently high, can affect the nearby community.

• *Research, development, testing and evaluation.* These types of facilities contain noise sources which could affect both personnel and communities. Examples include wind tunnels, machinery, and the testing of weapons, air/surface vehicles, and engines.

Source of Effects. The sources of noise which affect this attribute include:

• *Weapons.* Missiles and artillery of all types, including small arms, have extremely high noise levels and can severely affect the hearing threshold of their operators.

• *Vehicles.* Vehicles in the air, on the ground, or on water are important noise sources which affect the operator, other personnel, and the community. Examples include the following:

Aircrafts on and around commercial airports and military air bases significantly affect the community; particularly sonic booms or night operation, which can affect sleep patterns.

Large vehicles such as trucks, buses, tanks, and armored personnel carriers can affect the hearing threshold of the operators and passengers.

Most vehicles, when operated at night, can affect sleep patterns.

• *Construction equipment.* These types of equipment, which include vehicles and power tools, have high noise levels which can affect hearing thresholds of operators and site personnel.

• *Machinery.* Machinery in industrial plants, where noise levels are high and continuous, can significantly affect operator hearing thresholds.

Variables to be Measured. The important variables of noise which affect this attribute are its loudness, duration, and frequency content. As the loudness and/or duration increase, the effects of noise on the body increases. The internal bodily systems are increasingly under stress, the hearing threshold increases to the point where permanent damage (called noise-induced hearing loss) can occur, and sleep becomes increasingly impossible. Noises which contain high frequencies or contain, or are, pure tones, are more damaging and disturbing than those which do not.

Finally, the impulsivity of a noise is important. An impulsive noise is highly intense and short in duration (generally less than 1000 μsec), e.g., artillery or small arms fire. An impulsive noise is particularly important because it causes severe change in hearing threshold and can, if very intense, cause mechanical injury to the ear.[2,3]

How Variables are Measured. The loudness of noise is measured in terms of decibels (abbreviated as dB). Decibels are measured by using a sound-level meter. Normally, loudness is measured with a sound-level meter incorporating an "A" weighted electronic network.* The resulting measure is called dBA or dB(A). Most of the evaluation criteria are given in dB(A) units. The intensity of an impulsive noise may be different to read visually from a sound-level meter alone, due to the very short duration of the noise. To determine the intensity of an impulsive noise, refer the problem to acoustic engineers or to noise specialists.

The duration of a noise can be measured with recording equipment, a timer, or a stop watch. Impulsive-type noises can be expected to be less than 1 millisecond; measurement would require special equipment and measurement techniques.

The frequency content of noise is more difficult to measure, and complex equipment are required. Subjectively, however, high frequency content and pure tones are recognizable (assuming the observer's hearing is normal). For example, noises with high frequencies have a whine (jet aircraft), a screech (certain machinery), a clank or clink, a squeal, squeak, whistle, whine, or ping, or simply a tone. Noises with these characteristics are more annoying and disturbing to people and, at high loudness levels, more damaging. In general, subjective evaluations are not acceptable except to support or verify objective measurements.

Measurement for the physiological attribute, for existing situations, should be taken at the expected position of the human body with respect to the noise source(s). The sound-level meter should be placed where the body or people are or will be located. When the noise source is active, several readings should be taken and averaged (see discussion in Geographical and Temporal Limitations section). For accurate and precise measurement refer to *Handbook of Noise Measurement*, by General Radio Company[8], and/or *Methods for the Measurement of Sound Pressure Levels*, by American National Standards Institute (ANSI S1.13-1971).[9]

In those situations where the noise source is in the future and thus cannot be measured directly, either:

Measure comparable, existing situations.

---

*A network that closely responds to the response of the human ear and estimates damage potential.

Refer to published documents containing approximate dB(A) or dB levels and durations of comparable noise sources.*

Refer the problem to acoustics engineers or noise specialists. This alternative is especially recommended in complex situations or where large numbers of noise sources are involved.

Finally, to measure hearing loss or hearing thresholds of an individual, an audiometer should be used by a trained person certified as an audiometric technician, under the supervision of a physician or an audiologist. It is important to measure an individual's hearing before he is subjected to noise sources so that a baseline audiogram can be prepared. This audiogram can then be used for future references and comparisons with later tests.

Evaluation and Interpretation of Data.   The following criteria can be used to determine if the noise source will affect the body in any manner. The intensity and duration of the noise at the body should not exceed the values given below.

	Intensity	*Duration*
Internal bodily system	85dB (A) (6)	Any
Hearing threshold (continuous sound, if sound	80dB (A) (a)	16 hr
of intermittent summation is required; use	85dB (A)	8 hr
meters especially designed for this pur-	90dB (A)	4 hr
pose or contact audio engineers or sound	95dB (A)	2 hr
specialists)	100dB (A)	1 hr
	105dB (A)	30 min
	110dB (A)	15 min
	115dB (A)	7.5 min
	$>$ 115dB (A)	Never
Hearing threshold (b)		
(Impulsive sound)	140dB (at ear) (a)	1000 μsec
Sleep pattern		
(Causes awakening)	55–60dB (A)	Any
(Causes shift in sleep)	35–45dB (A)	Any

(a) American Conference of Government Industrial Hygienists (ACGIH), 1973.
(b) Operators of artillery and small arms can be expected to receive higher intensity levels.

*These dB(A) levels may be changed to reflect different distances between the noise source and measurements by applying the rule of subtracting or adding 6db(A) per doubing or halving of distance. For example, if the estimate is given as 90dB(A) measured at 50 feet and the actual distance between the noise source and the personnel is 100 feet, the dB(A) can be estimated to be 84dB(A). Noise sources which are "line" sources, such as trains and heavy streams of traffic, reduce in noise level 3dB(A) per doubling of distance.

If these values are exceeded, the noise source may harmfully affect the body. Should these values be exceeded in the community (even from air/ surface transportation), an environmental impact statement and further detailed measurements and evaluations will be required.*

Special Conditions.   The most serious noise impacts on this attribute are:

Partial hearing loss caused by artillery or small arms fire
Partial hearing loss to operators caused by combat, construction, or other vehicles
Sleep loss to the aged and the population in general or to recuperating patients caused by high noise levels and night operations.

Noise-induced hearing loss due to artillery, small arms fire, or combat vehicles is not uncommon in the military. Military personnel exposed to these noise sources should have their hearing checked periodically.

Noise sources of any type should not be located near hospitals or homes for the aged. Night operations should be isolated from these places and from any areas where people are sleeping.

Geographical and Temporal Limitations.   Any activities and noise sources should be geographically located so as to minimize their impact on communitites and populations. Isolation of the activity can be accomplished by geographic distance and/or placement with natural barriers (vegetation, hills, or mountains).

Noise sources affect people differently during the day. During the day people expect noise levels to be normal, but during the evenings when outdoor events, family activities, rest, television watching, etc., take place, noise levels are expected to be much less. At night, of course, noise sources are not expected to be active. Similarly, during weekends, noise sources should not be active.

In terms of measurement, variables should be measured or projected at various geographic distances and directions from the source until criterion values are reached to determine the extent of the noise. In addition, it is important to measure noise from transportation routes and flight patterns through communities and the air field. Also, the variables should be measured at various times during the day, evening, and night to determine the worst and best noise conditions.

Mitigation of Impact.   The optimal method of reducing sound level is, of course, to reduce the noise being produced by the source. Since this

---

*Assistance from audio engineers and audio specialists should be obtained.

method can be difficult or expensive to use on existing noise sources, the techniques of isolation and insulation are often used. If these techniques fail to reduce noise levels sufficiently, then the use of ear protective devices is recommended.

To reduce noise levels at the source requires engineering solutions. These solutions may include damping, absorption, dissipation, and deflection methods. Common techniques involve constructing sound enclosures, applying mufflers, mounting noise sources on isolators, and/or using materials with damping properties. Redesigning the mechanical operation of noise sources may be necessary. Performance specifications for noise represent a way to insure the procured item is controlled.

When an individual is exposed to steady noise levels above 85 dB(A), in spite of the efforts made to reduce noise level at the source, hearing conservation measures should be initiated.[12]

The federal government has promulgated three regulations that relate to controlling noise at the source. These noise regulations have been issued by the General Services Administration, the Environmental Protection Agency, and the Department of Labor.

General Services Administration. The General Services Administration issued construction-noise specifications effective July 1, 1972, for earth-moving, materials-handling, stationary, and impact equipment (see Table B-5). They require that all on-site equipment used by a contractor while under contract with the General Services Administration have A-weighted sound level requirements dB(A) measured 50 feet from the equipment. For example, a tractor, regardless of type, must not exceed 80 dB(A) while operation on the site at a distance of 50 feet. Noise violations result in a cancellation of the contract. Nearly all existing construction equipment exceeds these levels; therefore, some type of engineering noise control will be necessary.[1]

Environmental Protection Agency. Under provisions of the Noise Control Act of 1972, the Environmental Protection Agency is required to promulgate noise-emission standards for four new product categories:

Construction equipment
Transportation equipment
Motor or engine
Electrical or electronic equipment

In addition, all railroad and motor carriers engaged in interstate commerce will be subject to noise-emission requirements. Furthermore, any product adversely affecting the public health or welfare must be labelled with the specific sound level (see Noise Control Act of 1972). Although noise emission standards have not yet been issued, it is expected that engineering noise controls will have to be initiated.[1]

TABLE B-5. General Services Administration Construction-Noise Specifications

Equipment	Effective Dates	
	July 1, 1972	January 1, 1975
Earthmoving		
Front loader	79	75
Backhoes	85	75
Dozers	80	75
Tractors	80	75
Scrapers	88	80
Graders	85	75
Trucks	91	75
Pavers	89	80
Materials Handling		
Concrete mixers	85	75
Concrete pumps	82	75
Cranes	83	75
Derricks	88	75
Stationary		
Pumps	76	75
Generators	78	75
Compressors	81	75
Impact		
Pile drivers	101	95
Jack hammers	88	75
Rock drills	98	80
Pneumatic tools	86	80
Other		
Saws	78	75
Vibrators	76	75

Note: Equipment to be employed on this site shall not produce a noise level exceeding the following limits of dB(A) at a distance of 50 feet from the equipment under test in conformity with the Standards and Recommended Practices established by the Society of Automotive Engineers, Inc., including SAE Standard J 952 and SAE Recommended Practice J 184.

Department of Labor. Noise exposure criteria have been established by the Department of Labor under provisions of the Occupational Safety and Health Act. To meet the provisions of this Act, a hearing conservation program must be initiated for protecting noise-exposed personnel; emphasis should be placed on engineering noise control. Hearing protective devices should be issued to the workers, but only as an interim measure while engineering solutions are being planned.[1]

Other mitigation methods include isolation and insulation. The noise source and personnel or structures can be isolated from one another by distance.* (The intensity of noise decreases at an approximate rate of 6dB per doubling of distance.) Another method is to build barriers between the noise source and personnel, e.g., earthen barriers; walls of wood, cement, or block; or walls of trees and shrubs. Increasing insulation in structures will also reduce inside noise levels.

Finally, the predominant method of noise control has been the use of ear protective devices. Occupational health programs have emphasized the proper fitting and issuing of hearing protective devices** (i.e., ear plugs or ear muffs) to noise exposed personnel as an essential element of a hearing conservation program.

Secondary Effects. Exposure to high noise levels appears to have potentially detrimental effects on worker performance and accident rates and absenteeism in industry. In addition, this exposure may be presumed to cause general stress.[13] Continued noise production can lead to land use changes, with associated socioeconomic and biophysical ramifications.

Psychological Effects

Definition of the Attribute. Noise can affect an individual's mental stability and psychological response (annoyance, anxiety, fear, etc.).

Mental stability refers to the individual's ablty to mentally function or act in a normal manner. The mental well-being of an individual is essential for personal maintenance and efficiency. It is generally agreed that noise does not cause mental illness, but may aggravate existing mental or behavioral problems.[2,6] Noise predominantly causes psychological responses such as anger, irritability, increased nervousness, and most of all, annoyance. It is the annoyance reaction which can cause individual and community outcry and lawsuits against noise sources such as airports, aircraft, and highway transportation.

Activities that Affect the Attribute. Many activities can cause annoying and unacceptable noise. The most serious are:

• *Construction.* Construction projects create noise through the use of vehicles, construction equipment, and power tools. The noise affects operators, personnel, and communities near the site, and those people near transportation routes to the site.

---

*However, facilities should not be isolated to the extent that there is difficulty in getting to the facility or there is a possible reduction in use or usefulness of the facility.
**In situations where face-to-face communication is critical, hearing protective devices should include telephonic communication devices.

• *Operational activites.* The operation of most types of air/surface vehicles, machinery, power generating equipment, and weapons will generate noise. Maintenance and repair produce noise through the use of all types of tools, and when a number of noise sources are operating at the same time in the same general areas (e.g., a vehicle repair shop).

• *Military training.* Training courses and exercises which use any type of vehicle, weapon, power tools, appliances, or machinery, create noise for operators, military and civilian personnel, and, in large scale exercises, can affect nearby civilian communities.

• *Industrial plants.* The machinery and tools contained in these plants are a significant source of noise to the personnel, and, if noise levels are sufficiently high, can affect nearby communities.

• *Research, development, testing, and evaluation.* These types of facilities contain noise sources which could affect both personnel and communities. Examples include wind tunnels, machinery, experimental apparatus, and the testing of weapons, air/surface vehicles, and engines.

Source of Effects. The sources of noise which affect this attribute include:

• *Weapons.* Missiles and artillery, including small arms, have extremely high noise levels which can disturb and annoy personnel and nearby communities.

• *Vehicles.* Vehicles in the air, on the ground, or on water are significant sources of noise which annoy and disturb operators, personnel, and nearby communities. In particular, aircraft around airports and military bases can disturb and annoy base personnel and communities. Some individuals living directly beneath flight paths experience anxiety and fear from the aircraft noise. Additionally, they may find they must stop their work and mental processes due to passing aircraft, which, in turn, produces annoyance reactions.

• *Construction equipment.* These types of equipment which include vehicles and power tools have high noise levels which annoy operators, workers, and nearby community citizens.

Variables to be Measured. The important variables of noise which affect this attribute are its loudness, duration, and frequency content. As loudness and duration increase, psychological stress, annoyance, anger, and irritability also increase. In terms of frequency content, people are generally more annoyed by high frequencies and pure tones. The frequency content of a noise source also gives the sound an identity. Certain noises are annoying, disturbing, or fear-producing (to some people) because of their identity; e.g., sirens, jack hammers, horns, motorcycles, aircraft, buzzers, trucks, backfires, gunshots, and air compressors.

Noises which have very high noise levels, but very short durations (called impulsive noises), such as gunshorts, vehicle backfires, and sonic booms, startle people. These individuals are not only annoyed, but express feelings of fear and anxiety, and their activities (particularly sleep) are severely interrupted.

How Variables are Measured. The measurement of loudness, duration, frequency content, and impulsivity are discussed under *Physiological Effects*.

Evaluation and Interpretation of Data. It is difficult to establish a single set of criteria, due to the variety of acoustical and social factors. In addition to the intensity or loudness and duration of noise, other acoustical considerations involve pattern, occurrence, and the noise source itself. Social variables, such as demographic characteristics, personality type, and predisposition to nervousness, must be considered.[1]

While the spectral content and temporal patterns of noise pressure levels are important, as general criteria, ambient noise levels exceeding 50 to 55 dB(A) during the day or 45 to 55 db(A) during the night will disturb and annoy most people.*

Special Conditions. While environmental noise alone probably does not produce mental illness, the continual bombardment of noise on an already depressed or ill person cannot be helpful. Certainly it interferes with sleep, producing irritability and other tensions. Definite research has not been done in this area, but one 1969 study in England provides strong supporting evidence. Comparative studies of persons living adjacent to London's Heathrow Airport with others living in a quieter environment revealed that among those living in the noise environment there was a significantly higher rate of admission to mental hospitals.[6]

Another recent medical discovery is the effect of noise on unborn babies. Previously, unborn babies were thought to be insulated from the noise stress of the outside world, but now physicians believe that external noises can trigger changes in fetuses.[6]

Study of steelworkers indicates that those working in a noisy environment are more aggressive, distrustful, and irritable than workers in a quieter environment.[6] These studies show that it is very important to keep noise levels as low as possible in communities near hospitals, mental institutions, homes for the aged, and any place where people may be particularly annoyed or placed under mental stress by noise.

Geographical and Temporal Limitations. See discussion under *Physiological Effects*.

---

*These criteria are in agreement with most zoning ordinances in the United States.

Mitigation of Impact. Mitigation procedures relevant to the attribute are discussed under *Physiological Effects.*

Secondary Effects. Continued noise production can lead to land use changes, with the associated socioeconomic and biophysical ramifications.

Communication Effects

Definition of the Attribute. Noise can affect face-to-face and telephonic communication and, during extremely high levels of intensity, visual impairment has been reported.

Aural face-to-face communication, or the ability to give and receive information, signals, messages or commands, without instrumentation, is an essential activity. The temporary interference or interruption of communication during phases of human activity can be annoying, and occasionally hazardous, to personal well-being. Interference occurs when background or ambient noise levels of the environment are of sufficient intensity to mask speech, making it inaudible or unintelligible. Noise that interferes with communication can be dangerous, particularly when a message intended to alert a person to danger is masked, or when a command is not heard or understood. More commonly, however, noise is annoying because it disrupts the communication process.[*,1-3]

Telephonic communication, or the ability to give and receive information through telephones, headsets, receivers, etc., is also an important activity. Noise affects this type of communication in the same way as face-to-face, i.e., it causes annoyance and disruption. However, due to the insulation effect of the telephone or headsets and control over the volume of the incoming or outgoing signals, higher levels of loudness or intensity can be tolerated.[1]

Activities that Affect the Attribute. Many activities generate noise sufficient to interfere with aural communication:

• *Construction.* Construction projects create noise through the use of vehicles, construction equipment, and power tools. Noise levels are high enough to impact all types of communication, particularly for the operator and personnel in the general construction area.

• *Operational activities.* The operation of most types of vehicles, machinery, and weapons will create noise at levels which will interfere with communication of operators, personnel in the area, and communities. Com-

*One of the largest complaints in the community is that noise sources disrupt TV and radio listening.

munication in and near operating maintenance and repair shops will also be affected by the noise generated by tools and vehicles.

• *Military training.* Training exercises that use air, land and water vehicles, weapons, and machinery create noise levels sufficient to interfere with communication between military personnel.

• *Industrial plant activities.* Machinery and power tools contained in industrial plants are a significant source of noise affecting communication within the plant.

• *Research, development, testing, and evaluation.* These types of facilities contain noise sources which can affect communication. Examples include wind tunnels, machinery, experimental apparatus, and the testing of weapons, air/surface vehicles, engines, etc.

Source of Effects. The source of noise which affect this attribute include:

• *Weapons.* Weapons of all types (including small arms) have extremely high noise levels and can interrupt face-to-face communication. During large-scale activities, even telephonic communication can become difficult. Weapons achieve noise levels which may be sufficient to momentarily distort vision.

• *Vehicles.* Vehicles in the air, on the ground, or on water are important noise sources that affect the communication of operators and the community. Examples include the following:

Aircraft on and around airports and military air bases significantly affect communication, particularly in airport operations and in community areas directly beneath flight paths.

Large vehicles generate very high noise levels and can affect communication between the operators and other personnel in operating areas.

Other vehicles, particularly when operating in groups, affect communication near highways and other routes.

• *Construction equipment.* This equipment also has high noise levels and affects the intelligibility of communication at the construction site. Transportation or routes to the site may also generate noise levels that interfere with communication near the routes.

• *Machinery.* Machinery located in industrial plants where many machines are operating continuously can severely affect communication within the plant.

Variables to be Measured. The important variables of noise which affect face-to-face communication are loudness of the ambient noise level and the distance between the speaker and the listener. As the loudness increases, masking of the speech increases, and speech intelligibility and discriminability decrease. Also, as the distance between speaker and listener increases, the speech becomes more difficult to hear and to understand.

In telephonic communication, the noise variable of concern is the loudness of the background noise level.

As these variables increase, the speaker raises his voice to overcome the masking. Of course, the voice reaches a point where it strains and cannot overcome masking, and communication becomes impossible. In addition, the strain of shouting—and of trying to hear—is both fatiguing and frustrating in any situation, and may lead to inefficiency.[2]

How Variables are Measured.   The variable loudness can be measured or projected in dB(A) units, as specified in the *Physiological Effects* section. The distance between the speaker and receiver should be measured in feet.

Evaluation and Interpretation of Data.   The impact of noise on face-to-face communication can be evaluated by using the chart in Fig. B-6. Enter the side of the chart at the expected dB(A) noise level and the bottom at the expected average distance between speaker and listener. If the intersection of the two values falls above the Area of Nearly Normal Speech Communication, then speech communication is being adversely affected.

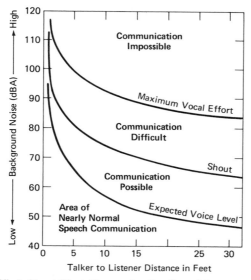

Fig. B-6.   **Simplified Chart That Shows the Quality of Speech Communication in Relation to Sound Level of Noise (dBA) and the Distance Between the Speaker and the Listener**[2]

In one-to-one personal conversation, the distance from speaker to listener is usually about 5 feet; nearly normal speech communication can proceed in noise levels as high as 66 dB(A). Many conversations involve groups; for this situation, distances of 5 to 12 feet are common and the intensity level of the background noise should be less than 50 to 60 dB(A). At public meetings, outdoor training sessions, or at construction sites, distances between speaker and listener are often about 12 to 30 feet, and the sound level of the background noise should be kept below 45 to 55 dB(A) if nearly normal speech communication is to be possible.[2]

In telephonic communication, background intensity levels above 65 dB become increasingly intrusive (see Fig. B-7).

Special Conditions.   There are special areas where communication which should not be disturbed takes place. These areas include training and testing areas, schools, churches, libraries, theaters, military communications centers, offices, hospitals, and research laboratories. Noise sources including air and land transportation routes should be isolated from these communication-sensitive areas or the areas should be well insulated against external noise.

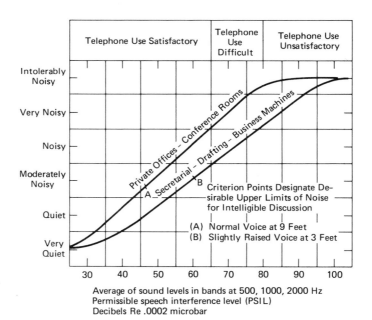

Fig. B-7.   Rating Chart for Office Noises.[8] Data were determined by an octave-band analysis and correlated with subjective tests.

Geographical and Temporal Limitations. See discussion under *Physiological Effects.*

Mitigation of Impacts. To insure intelligible communication, the noise sources and the personnel need to be isolated or insulated from one another. The special areas (see *Special Conditions*), where communication is especially sensitive, should be well isolated and insulated against external noise.

When it is unavoidable to have personnel who must communicate near high noise levels, communication devices should be used (e.g., headsets, field telephones).

Secondary Effects. As communication becomes more and more difficult to accomplish, impacts on psychological and performance effects may occur. Continued difficulties may lead to land use changes, precipitating a series of socioeconomic and biophysical ramifications.

Performance Effects

Definition of the Attribute. Noise can affect the ability of the human to perform mechanical and mental tasks. Noise can adversely affect performance through:

The increase in muscular tension that can interfere with movement

The lapse in attention or a diversion of attention from the task at hand

The masking of needed auditory signals

The startle response to high intensity noises.

Mechanical tasks can range from simple mechanical assembly to more complex tasks. Lower order tasks, such as mechanical assembly or manual routine-type activity, are least influenced by noise. However, tasks of this nature are altered in three essential ways by high intensity noise. Although work output remains fairly constant, worker errors can increase[2], judgment of time intervals can become distorted, and a greater effort is necessary to remain alert.[1,4] Noise is most likely to affect the performance of tasks which are quite demanding and/or require constant alertness.[3]

Mental tasks, such as problem solving and creative thinking, are more affected by noise. Higher order tasks requiring greater mental facilities (although dependent on the individual) are generally disrupted by lower noise intensities than mechanical tasks. It is important, therefore, to keep noise at a minimum in and near office areas.[3]

When a task (mental or mechanical) requires the use of auditory signals, speech or nonspeech, noise at any intensity level sufficient to mask or interfere with the perception of these signals will interfere with the performance of the task.[2]

Activities that Affect the Attribute. The most important activities that can reach noise levels sufficient to affect performance are:

• *Construction.* Construction projects create noise through the use of vehicles, construction equipment, and power tools. The noise levels are certainly high enough to affect mental tasks. The mechanical tasks at the site are generally considered highly physical and probably would be unaffected by the noise levels.

• *Operational activities.* The operation of all types of vehicles, machinery, and weapons will create noise levels sufficient to affect human performance, primarily through distraction.

• *Military training.* Training exercises which use vehicles, weapons, and machinery create noise levels high enough to distract people.

• *Industrial plant activity.* Machinery and power tools contained in industrial plants are a significant source of noise which affects mental and mechanical performance.

• *Research, development, testing, and evaluation.* These types of facilities contain noise sources that can affect performance. Examples include: wind tunnels, machinery, experimental apparatus, and testing of weapons, air/surface vehicles, and engines.

Source of Effects. The sources of noise that can affect performance include:

• *Weapons.* Weapons of all types generate noise levels which can interfere and interrupt mental and complex, precise mechanical tasks.

• *Vehicles.* Vehicles of all types are significant noise sources which can interfere with and interrupt task performance. In particular, aircraft can disrupt the mental tasks of large segments of communities.

• *Machinery.* Machinery and power tools in industrial areas create high noise levels that could affect some complex and precise mechanical tasks.

• *Construction equipment.* Noise from this equipment can affect the mental tasks of personnel in the area.

Variables to be Measured. The important variable of noise which affects task performance is loudness. As the loudness of noise increases, the effects of the noise on performance increase. First, mental tasks are affected, then, as the loudness further increases, complex and precise mechanical tasks become affected.

How Variables are Measured.   Loudness of noise is measured in terms of decibels.   A detailed discussion of how to measure or project decibel levels can be found under *Physiological Effects.*

Evaluation and Interpretation of Data.   The following criteria can be used to determine if the noise source will affect task performance.   If the intensity of the noise source exceeds the values given below, task performance may be affected, depending upon the spectral content and temporal patterns of the noise.

Task	*Intensity*
Mechanical, manual, or mental repetitious	85 dB(A)
Mental (problem solving or creative)	See Table B-6

Special Conditions.   Special areas where mental tasks take place should not be disturbed.   These areas include offices, conference areas, schools, indoor training areas, libraries, and research laboratories.   In terms of mechanical tasks, it is difficult to be specific.   Wherever complex, precise, and demanding mechanical tasks are performed, the environment should be protected from high intensity noise sources.

TABLE B-6.   Noise Criteria for Mental Tasks

Type of Room	Maximum Permissible Level (measured when room is not in use)
Small private office	45 dB(A)
Conference room	35–40 dB(A)
Secretarial offices (typing)	60 dB(A)
School rooms	30–35 dB(A)
Reading	40–45 dB(A)
Meditation	40 dB(A)
Studying	40–45 dB(A)
Individual creative activity	40–45 dB(A)

*These figures are general guidelines. For specification purposes, Noise Criterion (NC) curves which cover the different frequency bands should be used (see Ref 14).

Geographical and Temporal Limitations. See discussion under *Physiological Effects.*

Mitigation of Impact. The optimal method of reducing sound levels is through noise source reduction, isolation, or insulation. Methods of achieving noise reduction are discussed under *Physiological Effects.*

Secondary Effects. Continued exposure to noise may lead to and land use changes, precipitating a series of socioeconomic and biophysical impacts.

Social Behavior Effects

Definition of the Attribute. Social behavior refers to the individual's ability to mentally function in a normal manner on an interpersonal basis. Under certain conditions within communities, interpersonal relationships are altered when noise is of sufficient intensity. Areas of socialization may become restricted due to noise exposure. Outdoor areas are first to be affected, thus limiting socializaton to residential interiors. Patterns of entertainment become confined and restricted. When one or more methods of basic auditory communication (face-to-face or telephonic) are masked, the channels for social interaction become limited. These results, in turn, affect personal attitudes and create annoyance.

Activities that Affect the Attribute. Many activities generate noise levels which could interfere with social behavior.

• *Construction.* Construction projects generate sufficient noise to interfere with the social behavior of personnel and communities located near the site. In particular, new transportation routes to the site will introduce new noise levels to people living nearby which, in turn, can adversely affect social behavior.

• *Operational activities.* The operation of air/surface vehicles and other equipment increases the noise level of the outside environment, where much socialization takes place. Social behavior inside structures can also be affected by these activities, if their noise levels are extreme.

• *Military training.* Training exercises which use vehicles, weapons, and machinery create noise levels sufficient to interfere with socialization between personnel. Noise from large scale training exercises can also affect the community.

• *Industrial plant.* Machinery and activities in the plant are a significant source of noise which affect social behavior between personnel and in nearby communities.

• *Research, development, testing, and evaluation.* These types of facilities contain noise sources which can affect socialization. Examples include wind tunnels, machinery, experimental apparatus, and testing of weapons, air/surface vehicles, and engines.

Source of Effects. Most noise sources are capable of influencing social behavior in community environments, particularly outdoors. In and around military air bases and airports, weapons, construction equipment, aircraft, helicopters, and ground-transportation vehicles generate noise levels sufficient to interfere with interpersonal communication, thus affecting and limiting social behavior on the base and in surrounding communities. Social behavior inside structures is probably affected mostly by machines, heating and air-conditioning units, operating appliances, research equipment, or external noise sources with very high levels such as aircraft, trucks, and construction equipment.

Variables to be Measured. The important variables affecting this attribute are the same as those discussed under Communication Effects. As communication becomes difficult or impossible, social behavior and interpersonal relationships become limited, especially in the outdoor environment.

How Variables are Measured. See discussion under *Communication Effects* and *Physiological Effects*.

Evaluation and Interpretation of Data. Evaluation techniques, as outlined under *Communication Effects* and *Psychological Effects*, should be applied to this attribute for both outdoor and indoor environments.

Special Conditions. Social behavior is important to people. Being able to socialize with friends, neighbors, and members of the family is an essential human activity. Constant interruption of these activities can only result in frustration and annoyance. Consequently, it is important to consider the impact of continuous or highly repetitive noise sources such as aircraft, weapons, and vehicles on social behavior in surrounding communities.

Geographical and Temporal Limitation. Socialization is normally expected to take place in the community, and in and around living and entertainment areas. These areas then should be measured (inside and out) to determine if any noise sources are affecting social behaviors. One of the most prevalent noise sources is from transportation, both from routes through communities and flight paths. These, too, must be measured. Socialization occurs most often in the evening, early night hours, and on weekends. Measurement of noise during these periods should be emphasized for this attribute.

Mitigation of Impact. Social behavior is primarily affected by noise sources which create high noise levels in the outdoor environment. The mitigation techniques of source control, isolation of the sources from the community, or creation of barriers would be useful. Discussion of these techniques is found under *Physiological Effects.*

Secondary Effects. Continued exposure to noise may lead to land use changes, precipitating a series of socioeconomic and biophysical impacts.

REFERENCES

1. Jain, R. K, *et al.*, "Environmental Impact Study for Army Military Programs," *Technical Report No. IRD 13*, U.S. Army Construction Engineering Research Laboratory, Champaign, Illinois, December 1973.

2. Miller, J. D., *Effects of Noise on People*, U.S. Environmental Protection Agency, Washington, D.C., December 31, 1971, PB-206 723.

3. Panel on Noise Abatement, *The Noise Around Us*, U.S. Department of Commerce, September 1970, COM 71-00147.

4. Kryter, K. D., *The Effects of Noise on Man*, Academic Press, New York, 1970.

5. National Bureau of Standards, *The Social Impact of Noise*, U.S. Environmental Protection Agency, December 31, 1971, PB-206 724.

6. U.S. Environmental Protection Agency, *Noise Pollution*, August 1972, U.S. Government Printing Office, Washington, D.C., 5500-0072.

7. U.S. Environmental Protection Agency, *Report to the President and Congress on Noise*, February 1972, U.S. Government Printing Office, Washington, D.C., 5500-0040.

8. Peterson, A. P. G. and E. E. Gross, *Handbook of Noise Measurement*, General Radio Company, West Concord, Massachusetts, 1972.

9. American National Standard, *Methods for Measurement of Sound Pressure Levels*, American National Standards Institute, New York, ANSI S1.13-1971.

10. Wilsey and Ham, *Aircraft Noise Impact Planning Guidelines for Local Agencies*, U.S. Department of Housing and Urban Development, November 1972, NTIS-PB-213020.

11. U.S. Army Military Standard, *Noise Limits for Army Materiel*, MIL-STD-1474 (MI), 1 March 1973.

12. "Noise and Conservation of Hearing," U.S. Army, *Technical Bulletin Med.* **251**, March 1972.

13. U.S. Environmental Protection Agency, *Public Health and Welfare Criteria for Noise*, July 27, 1973.

14. Beranek, L.L., *Noise Reduction,* McGraw-Hill, New York, 1960.

HUMAN ASPECTS

A critical aspect of man's environment is characterized by the way in which man interacts with other men and the natural environment. Owing to the complexity of his activities and interrelationships, it is difficult to identify general parameters that describe the condition of human resources. The attributes that have been identified for this purpose are obviously not completely descriptive of all man's activities and may appear to miss many important issues. Nevertheless, these few attributes have been chosen because, for most projects envisioned, they will be able, if applied, to capture the major human (or community) elements of environmental impact.

Because of their generality, the attributes are difficult to measure and define, and as a general rule, an adequate assessment of impact on human resources will probably have to be undertaken by persons with special expertise in this area (sociologists or psychologists). The attributes that should be examined include:

Life styles
Psychological needs
Physiological needs
Community needs

Life Styles

Definition of the Attribute. This attribute refers to the many social activities of man. Such activities often take on structural characteristics which eventually cause them to be organizations. The makeup of organizations may vary, depending upon the characteristics, interests, and objectives of the organization population. Some bases for these organizations could be racial, ethnic, political, religious, and occupational. Another perspective of this attribute occurs in the form of informal interaction between friends, relatives, and co-workers.

Activities that Affect the Attribute. The major classifications of activities affecting this attribute include those which affect employment and job security, standards of living,\community development,\and recreational op-portunity. Examples of these include population migration, transporta-tion projects, large construction projects, water resources projects, and industrially-related activities. A number of minor activities falling within these major categories could affect this attribute.

Source of Effects. Change or impacts that occur in this attribute will be dependent upon changes that occur in the population. For example, in a community where various activities have been established, the outmigra-tion of a large portion of the population could disrupt a number of both formal and informal activities. Examples of some of these activities are athletic groups, schools, and church groups. Individual, informal inter-actions are also to be considered in this attribute.

The following example is given to illustrate how a significant change in population can cause changes in this attribute. If the population were pre-dominantly elderly people, the type of activities they would be involved in might include hobby clubs, craft clubs, and card-playing clubs. If a large portion of this population were to migrate out of the area, the stability of some of these activities might be affected. Likewise, if many more people of the same age group, with the same interests, moved into the community, the stability of these groups might be strengthened. Also, if the population mix were changed significantly, perhaps by an influx of many more younger people, the stability of such groups might be threatened.

Variables to be Measured. The variables to be measured for this attribute cannot be precisely identified. The purpose in considering this attribute is to identify those instances in which a noticeable change will occur that will affect many people. The objective is to identify general changes in social activities and practices which will be caused by the proposed action.

How Variables are Measured. The variables in this attribute cannot be precisely measured. One approach that can be taken to "measure" changes in this attribute would involve making a survey of the area to determine the number and kinds of social organizations and activities that exist before the proposed action takes place. Then, having determined the changes ex-pected to occur in the population and the characteristics of that population, impacts on organizations and activities can be predicted in terms of how they may grow, desist, or experience noticeable alterations.

Some persons who might be good sources for predicting and interpreting impacts in this attribute are leaders and participants in the organizations, coaches of local athletic teams, community social and recreation leaders, and local political leaders.

Evaluation and Interpretation of Data. Interpretation of the impacts or changes in this attribute must be performed by the impact assessment, which should then be analyzed in conjunction with the opinions and assessment of the people mentioned above.

Geographical and Temporal Limitations. Usually, a geographic area larger than that of the immediate community must be included in the analysis, because impacts often occur outside of the area following changes within the community. The residence and work locations of those who take part in the activities that are considered a part of this attribute must be included in the geographic area.

The analysis should include a summarization of this attribute before the proposed action takes place and the anticipated changes that will result from the proposed activity. In addition to these considerations, to get a more realistic view of the total impacts, consideration must be given to what the condition of this attribute would be if the present situation, or the previous situation, were to continue, as well as to what changes would occur normally, without the proposed action.

Mitigation of Impact. Although impacts to this attribute cannot be completely mitigated (with the exception of postponing the proposed action indefinitely), the effect of anticipated impacts could be lessened simply by forewarning participants that such changes are expected to occur. This will enable the organizations and participants of informal activities to prepare themselves to adjust to expected impacts.

Secondary Effects. Socioeconomic changes frequently may result in secondary or indirect impacts on the biophysical environment. These impacts need to be identified and assessed.

In the case of social changes, there may be environmental effects on air, water, land, etc., from increasing or decreasing the population in an area. Additional people cause increased demand for water, sewage treatment, and power; they require new housing, which takes land, shopping centers, and schools; they require transportation, which increases traffic congestion and degrades air quality.

Psychological Needs

Definition of the Attribute. This attribute refers to the needs of human beings that can be distinguished from the physiological needs, primarily those of emotional stability and security. Although this attribute could relate to such factors as instincts, learning processes, motivation, and behavior, these factors are not included in this attribute because of the difficulty in relating changes in outside factors to changes in these factors. Emotional stability and security are, therefore, the only two psychological needs that are considered in this attribute.

Activities that Affect the Attribute.   The major classifications of activities that are likely to affect this attribute are essentially the same as those affecting *Life Styles.*

Source of Effects.   Changes in the degree of emotional stability and feelings of security within the individuals affected could occur from a number of activities. For example, in construction or industrial activities, it may be necessary for some people to be moved from their homes or businesses. Even though it is difficult to anticipate the effect of such relocation, experience has shown that such activities may have negative effects on the people involved. These effects will vary in their degree of permanency.

Also, when the proposed activity would involve increasing or decreasing the number of jobs or other opportunities (e.g., recreational) in an area, it can be assumed that such activities will either increase or decrease feelings of security, particularly for those who are directly affected by the change in job availability.

Feelings of concern for physical security may be affected by fear for personal safety from crime elements, or from natural or man-induced disasters (e.g., nuclear power plants or industrial facilities).

Variables to be Measured.   Although no specific variables are identified for this attribute, a general feeling of the degree to which the psychological needs of individuals and communities are being met can be obtained.

Evaluation and Interpretation of Data; How Variables are Measured.   Data concerning impacts of this attribute must be obtained from several sources. One source would be detailed plans of the proposed activities and identification of groups who might be affected in such ways as in the examples above. This information could then be given to psychologists, who could best anticipate and interpret such changes that will occur as a result of the proposed activity. Impacts in this attribute cannot be measured, but can be identified as to whether the impacts are potentially beneficial or disruptive.

Other information may be obtained from personal surveys, or by consulting local counselors, clergy, and law enforcement officials.

Geographical and Temporal Limitations.   The geographic area for this attribute must be that which contains people who would be affected by it.

This, therefore, could include areas both within and outside of the immediate community.

The time limits for this attribute would the the same as for the other attributes in this section. The "before" time period should be that time shortly before instigation of the proposed activity. The "after" time period should include that time immediately after the proposed activity has been completed.

Mitigation of Impact.   Some adverse impacts might be averted by including, in the proposed activity funds, an action plan that would permit assistance for those people who would be impacted. For example, when a number of jobs are to be disbanded, a service could be set up in which those people who would be without jobs could obtain assistance in locating jobs in other areas. In problems caused by relocation, some program of assistance could be instituted in which people could be aided in finding housing and business locations similar to those they now have.

Fears for personal safety may be alleviated through planned safety programs coordinated with public interest groups.

Secondary Effects.   See description under *Life Styles.*

Other Comments.   Even though impacts which may occur in this attribute are difficult to identify, measure, and evaluate, the attribute is included in the impact assessment process because it is very important. Therefore, it is necessary to attempt to identify situations where such impacts might occur, even if only the possibility of potential impacts can be identified, with very little interpretation or evaluation. This attribute is useful, at least in trying to anticipate where impacts may occur and in identifying situations for which mitigation procedures may have to be planned, and included in, the proposed activity.

Physiological Systems

Definition of the Attribute.   This attribute refers to anything that is a part of a person's body or that plays a part in a bodily function. It includes both individual parts (organs) and systems, such as the transport, respiratory, circulatory, digestive, skeletal, and excretory systems. All parts of the human body that contribute to its effective, efficient functioning are included in this attribute.

Activities that Affect the Attribute.   Major classifications of activities that can affect this attribute include construction; operational activities; military training and mission change; industrial; and research, development, testing, and evaluation. Any activity that can harm or threaten the efficient functioning of any part of the human body must be considered in light of its effect on this attribute.

Source of Effects.   The possible sources of impacts in this attribute are many. They range from activities performed in a laboratory to construction activities that might impair the safety of individuals working in the area. This attribute considers any hazards that may impair safety of any individual.

Variables to be Measured.   There is not a list of variables that can be measured for this attribute. The purpose of this attribute is to identify potential sources of harm to people. Therefore, detailed activities and implications of the proposed activity must be examined to determine if any of those activities may be potentially harmful.

How Variables are Measured; Evaluation and Interpretation of Data.   It would be helpful to rely upon the knowledge and skills of people who are familiar with the kinds of harm considered in this attribute that can occur. It is suggested that physicians be contacted and given a description of the proposed activity. The seriousness of the potential impacts can then be determined through professional opinion.

Special Conditions.   It must be determined how many persons will be affected by the expected impacts. Although the impacts are not considered slight even if they affect only a few, it may be said that seriousness will increase as the number of affected people increases.

Mitigation of Impact.   Anticipated impacts in this attribute can be mitigated by taking whatever precautionary measures are necessary to avoid the impact. This may take the form of including in the proposed activity specific safety practices and protective devices.

Secondary Effects.   Effects on physiological systems can also affect psychological needs, and may have additional economic ramifications if a significant number of workers or production is affected.

Community Needs

Definition of the Attribute.   This attribute refers to some of the many services that a community requires. It includes such things as housing; water supply; sewage disposal facilities; utilities such as gas, electricity, and telephone; recreational facilities; and police and fire protection. The nature of change or impact that occurs in this attribute as a result of the proposed activity will be very much dependent upon the type of change that is expected to occur in the population as a result of this proposed activity.

Activities that Affect the Attribute. Major classifications of activities that are likely to affect this attribute are essentially the same as those affecting *Life Styles.*

Source of Effects. As changes in population and characteristics of the population occur, the needs or services required for that population will change, too. For example, in the general activity category of construction, a temporary force of construction workers may be required to perform the activity. If the construction workers and their families must settle in an area until the construction is completed, these workers and their families will require particular services, such as those mentioned in this attribute. Likewise, when they leave the community, the demand for these services will have been lessened, or perhaps even dissolved, thus leaving the community with a supply of services that is no longer needed.

Also, in industrial activities, a number of people may be brought into an area on a permanent basis, and the community may find itself unprepared to provide the services and needs to this permanent addition to the population. Also, impacts can occur as a result of change in military mission or a change in the number of training activities taking place on a particular base. These impacts may result from fewer numbers of people requiring the services that have already been designed to serve a greater number of people. For example, a community may find itself with an oversupply of houses or have to decrease the number of personnel required for such activities as police and fire protection.

In these and other activities, there are particular sub-activities that relate directly to the provision of some of these services. Therefore, any proposed activity that has to do with the provision of such services should be investigated as to the impact that will occur.

Variables to be Measured. For the impact assessment procedure, variables that should be measured are those which will indicate services in the community that are available as well as what services are needed. The community should be surveyed in order to determine (1) the change in population and the characteristics of that population; (2) the number of houses and apartments available to meet the needs of the population if there will be an increase; (3) the number of homes supplied with water and sewage disposal facilities and other facilities; (4) the number of personnel on the police force and the fire department; and (5) the number of acres of land devoted to recreational activities and the number of recreational activities available in the area.

How Variables are Measured. Communities should be surveyed to determine what services are now available. For example, a survey should be made to determine the number of available dwelling units (houses, apartments, and trailers, for example) that are available and the number of those served with adequate water, sewage, and utility service. The availability of recreational facilities can be determined by noting the number of acres devoted to recreational usage and the number of recreational activities available. The number of police and fire protection personnel should be determined to indicate the level of service now available to the population.

Various sources can be utilized for obtaining this information. Planning agencies often have information on all of these services. Police and fire department personnel are sources which can give an indication of the adequacy of these kinds of services.

After this information is obtained, it will be necessary to relate the present conditions to the change in population that is anticipated from the proposed activity. If the population will increase, it must be determined if there are enough facilities and services available to serve the incoming population. On the other hand, if there will be a decrease in population or an outmigration, the services provided by the community must be considered in light of the oncoming decrease in demand. Perhaps other uses can be made of those services no longer in demand in their usual functions.

Evaluation and Interpretation of Data. There are no standardized means of interpreting the above mentioned variables. For the purposes of an impact assessment, when anticipated changes in population of an area will cause serious problems in the services needed by the population, the situation must be further studied for the impact statement. Expert judgment may be useful in determining when a serious problem will exist, given an immigration or outmigration of a significant number of people in the community.

Geographical and Temporal Limitations. The geographic area to be considered in this attribute will vary, depending upon the proposed activity. The area to be considered will depend upon where the affected population resides and works. Therefore, any area where people who will be affected by the services discussed herein reside or work must be considered in the determination of impact.

In determining the impact that occurs within this attribute, the analysis must be done for the area before and after the proposed activity is instituted.

It is suggested that the "before" time period incorporate those conditions that exist or can be anticipated to exist shortly before the proposed activity is instituted. It is also suggested that the "after" time period be that time period shortly after the proposed activity has been completed.

Mitigation of Impact. Impacts in this attribute can be mitigated by including in the planning process for the proposed activity a plan for providing the services that have been identified as being needed or proposing alternative uses that can be made of services that will no longer be needed as such by the population.

Secondary Effects. See description under *Life Styles.*

ECONOMICS

The potential impact on the economic structure of changes resulting from project activities stems primarily from the direct effect of purchases of goods and services for project activities and the indirect effects arising from goods and services purchased from payrolls. These effects may be summarized by reference to three major attributes that reflect impact on industrial and commercial activities, the local government, and the individual. These attributes are as follows:

Regional economic stability
Public sector revenue and expenditures
Per capita consumption

Regional Economic Stability

Definition of the Attribute. This attribute indicates a change in the ability of a region's economy to withstand severe fluctuations, or the speed and ease an economy demonstrates in returning to an equilibrium situation after receiving a shock. This is an ex post definition, whereas a surrogate, ex ante definition is the diversity of a regional economy or the degree of homogeneity of the region's economic activities in contributing to the gross regional product. The more diverse an economy and the more closely related it is to growth areas of the national economy, the more stable it is likely to be.

Activities that Affect the Attribute. Any activity that results in some input or output relationship with a local business or individual has an impact on the growth and stability of the regional economy. Direct purchases would have an effect, as would indirect purchases through payrolls.

Source of Effects.   The severity of a change in stability is directly proportional to the degree of dependence of the regional economy on the affected business for incomes and employment. Thus, if one or a few industries or firms dominate a region's economy (measured by the share of gross regional product or proportion of total employment), that region is highly sensitive to factors affecting those industries. Hence, activities that decrease the industrial diversity in an area are reducing the stability of the region, especially when the key industries are locally important and declining nationally.

Variables to be Measured.   Effects on the regional economy are indicated by the percentage of total regional economic activity affected by the activity. For example, if 25 percent of all retail sales in a county stem from agency personnel purchases, significant impacts can be anticipated from a change in personnel. Likewise, the agency's direct purchase of labor or other materials from the local economy should be examined as a percentage of local economic activity.

How Variables are Measured.   Considerable ingenuity must be exhibited by the individual who is measuring impact on regional economic stability. Variables to be examined would include employment in economic activity related to specific activities. Production and income variables might also be examined.

Evaluation and Interpretation of Data.   There are no rules that would enable one to determine whether or not a given change is small or large. Instead, judgment must be exercised with explicit reference to the basis for judgment. This approach would enable any reviewer to evaluate the facts and, perhaps, to disagree with the judgment. At least, full consideration of the issues and the rationale for a conclusion will have been given.

Special Conditions.   Stability and, perhaps, growth are two goals of a regional economy. They are usually incompatible because, in the long run, some specialization is required if a growth rate higher than that for the rest of the country is to be realized. Therefore, the unique or special characteristics of the regional economy must be considered. An economy with an agricultural base, for example, might be much more severely impacted by the withdrawal of agricultural land for use in a project than if agricultural land were to be withdrawn from use in an industrial-based economy.

Geographical and Temporal Limitations. The same geographic and temporal limitations that exist for the per capita consumption attribute are applicable here.

Mitigation of Impact. Mitigation of negative effects can be achieved in one of two ways: Either increasing the demand for the output of high growth industries in the region, or changing the distribution of demand for the output of different firms so that the resulting employment redistribution approximates more closely the situation at the national level (taking into account the potential for regional specialization).

Secondary Effects. Economic changes frequently result in secondary or indirect impacts on the biophysical environment. These impacts need to be identified and assessed.

In the case of economic effects, programs or actions that add or reduce revenue in an area will result in additional or decreased population and new economic activity in local communities. This may take the form of new or fewer retail outlets (stores, garages, etc.), increased or decreased service-oriented businesses, and land use changes as new home developments, shopping centers, etc., are created, or requirements for them are reduced. Most of these activities will have a secondary impact on air, water, and land attributes.

Public Sector Revenue and Expenditures

Definition of the Attribute. This attribute is the annual per capita revenues and expenditures of local and state governments and associated agencies in the region under study. Changes in this variable can be interpreted as a measure of the change in economic well-being of the public sector.

Activities that Affect the Attribute. Changes in the economic, social, or physical conditions of the area due to project activities may result in changes in public sector revenues and expenditures. The effects would be felt primarily through changes in employment, industrial or manufacturing activities, and the acquisition of release of real estate by agency action.

Source of Effects. Tax receipts are directly affected by changes in personal income. Payments from the federal government to local governments, to compensate for increased local expenses, also may occur. Changes in land usage and, therefore, assessed value also affects revenues collected.

Numerous changes in the costs of services (and therefore on requirements for public expenditures) occur in such areas as education, transportation, public welfare, health, utilities, and natural resources, as the direct result of an activity or indirectly through employment changes caused by the activity.

Variables to be Measured.   One measure of impact is the average annual revenues and expenditures of the relevant government and its agencies in a defined geographic region over the lifetime of the project, assuming the project or activity has been undertaken, minus the same measure over the same time span, but assuming the activity has not been undertaken (and everything else the same).   In lieu of the annual average, a particular year may be chosen arbitrarily and the change in annual net revenue computed for that year.

Another set of variables would be a comparison, on a function-by-function basis, of the expenditures necessary to provide adequate public services with and without the project.

How Variables are Measured.   The geographic extent of the impacted public sector must be defined a priori, usually as a local (town, city, or county) or state government.   Changes in revenues and expenditures must then be estimated on an item-by-item basis.   Tax revenue changes may be estimated as described in the section on measurement of variables in the per capita consumption attribute.   Local sales tax rates should be used in lieu of state rates where pertinent, and the state or local income tax rates should be used in place of the composite national rate as described in the per capita consumption attribute.   Effective state income tax rates can be found in *State and Local Finances*, in the table of "Effective Rates of State Personal Income Taxes for Selected Adjusted Gross Income Levels, Married Couple With Two Dependents, by State," where local rates are simply the given percentages (no deductions).[1]   Corporate tax receipts for local areas are generally not important.*

The change in gasoline tax receipts is determined by calculating the percentage change in the number of vehicles in the area.   This implies that the tax rate, the per mile gasoline consumption for each vehicle, and the total mileage per vehicle are constant.   The percentage change in vehicles may also be assumed to be proportional to the change in personal income.   Independent estimates may be made through interviewing automobile dealers

---

*Where the entire state is the pertinent impact area, corporate tax revenues can be determined if it is assumed that corporate profits change in the same proportion as labor costs and production.   The ratio of corporate to individual income tax receipts can be determined from the *Statistical Abstract of the United States*, in Table 4, "State Tax Collections and Excise Taxes, by Type of Tax—State."   Multiplying this ratio by the change in personal income calculated in the per capita consumption attribute results in an approximation of corporation taxes.

or by multiplying population changes times a factor representing cars per capita. The preliminary value for motor fuel tax receipts can be based on figures from the *Statistical Abstract of the United States*, in Table 4, "State Tax Collections and Excise Taxes, by Type of Tax—State," where the state receipts must be multiplied by some proportion to determine the local share (this proportion may depend upon the gasoline sales, and hence, indirectly depends on the personal income in the area). Local data should replace the extrapolated state data where available. Since the gasoline tax receipts for some future year (under the assumption that the activity has not been undertaken) is the basis for the measurement; at a minimum, tax receipts for at least two past years should be linearly extrapolated to arrive at the desired figure.

Changes in payments to the local government or its agencies by individuals, businesses, and other agencies, for particular goods or services (e.g., water and other public utilities) should be included on a specific basis. Transfer payments from outside sources that are direct or indirect compensations for incurred expenses should be based on the specific changes in these costs caused by the project activity, following standard reimbursement procedures. For example, compensation for increased educational expenses for military families is a transfer payment to the local area. Total changes in receipts from taxes, subsidies, and transfer payments due to the project activity should be summed to arrive at an aggregate figure.

Changes in local public expenses due to the activity may be assumed to be proportional to changes in the total personal income in the area, reflecting both the number of consumers of a public good or service and the per capita level of consumption. The *Statistical Abstract of the United States*, in the table of "Direct General Expenditure of State and Local Governments—State" gives figures for public sector expenditures for $1000 of personal income. These ratios must be multiplied by the proportion of expenses accruing to the local government for each category: education, highways, and health. When available, these ratios should be calculated from local information for all types of public goods and services that change in the same proportion as total personal income. Some expenses, such as welfare payments, do not change proportionally and must be calculated through independent analyses. Among these expenditures are damages to public facilities or any other temporary or permanent costs identified as resulting from the project activity, but not through social or economic changes within the population. The percentage change in personal income in the impact area is determined from the per capita consumption attribute, and this proportion must be multiplied by the public sector expenses per unit of income. The resulting figures are the changes in public expenditures if the project occurs, and summing them gives the total change in expenses.

Evaluation and Interpretation of Data.   The changes in public sector revenues and expenditures must be compared to determine whether or not there is a net gain or loss to the public sector subsequent to the project. The severity of the impact (either positive or negative) would remain a matter of individual judgment and would be partly subject to considerations of indebtedness of the community.

Special Conditions.   The measurement can be improved if a more accurate estimate of future revenue and expenditure levels without the project can be determined. A detailed analysis, perhaps using multiple regression techniques, would improve these projections as well as help identify and evaluate more precisely the causal relationships between public sector revenues and costs and the direct impacts of the proposed activity.

Georgaphical and Temporal Limitations.   In general, the same geographic and temporal limitations that exist for the per capita consumption attribute are applicable in this measurement situation. The geographic range of local governments and civilian public agencies with respect to both the revenues and expenditures must be determined in a manner similar to an analysis of the market and supply areas of a private sector business enterprise.

Mitigation of Impact.   A negative impact can be mitigated if the project activities are designed to either reduce costs to the local community (e.g., demands for public sector goods and services, physical or economic damages to existing infrastructure) or to increase the direct or indirect payments to the local government.

Secondary Effects.   See the description under *Regional Economic Stability.*

Per Capita Consumption

Definition of the Attribute.   Annual per capita personal consumption of goods and services by local citizens is per capita consumption. This variable can be interpreted as a direct measure of personal economic well-being.

Activities that Affect the Attribute.   Increases (or decreases) in local employment, industrial expansion (or reduction or deletion), and construction all have the potential for affecting per capita consumption.

Source of Effects.   A change in demand for local goods or services results in increased or decreased money available for purchase of goods and services (disposable income). As another example, disposable income, and therefore, consumption may be affected by a changed tax base resulting from project acquisition of formerly taxable land.

Variables to be Measured. The baseline measure is the average amount that will be spent in each future year throughout the life of the project by each resident of the affected area for goods or services meant for personal consumption, assuming the project has not been undertaken. The variable indicating change is that same calculation, but under the assumption that the project or activity has been undertaken (with everything else exactly the same) minus the baseline measure.

How Variables are Measured. Assuming that businesses are not at full capacity, a change in final output (in dollars) will be reflected in a change in all short-run costs, including labor wages. With constant returns to scale for inputs, the change in completely proportional, and output revenue and all costs will change proportionately to the change in production, based on the current ratio of these values. In addition to labor costs, profits which accrue to the owners of a business may change. A determination of how a profit change affects personal income in the region must be based on an individual analysis of each business, with consideration for the amount of profit per dollar of output and the location of the owners (where the changed income of non-local residents is not included). Thus, a coefficient for a particular industry may be determined, showing the ratio of local personal income (wages, salaries, profits) to the dollar output of the industry.

Changes in output due to project activity may be approximately determined by first noting all industries, firms, or individuals who supply some needed input to the activity, and the amount of this input in dollars. Included as inputs are such goods and services as local raw materials, retail goods and services bought by project personnel and their families, and contributions to local charities.

Changes in the inputs (which are the outputs of the supplying firms) must be calculated or estimated with as much accuracy as possible. Assuming a constant, linear production function (constant input mix), the change in a supplying firm's output can be approximated by first determining the ratio of the activity's current requirements for the firm's output (in dollar terms) to the current total requirements for that activity (which need not be measurable in dollar terms). Multiplying this ratio by the change in activity, the change in the firm's output is determined. This is multiplied by the previously calculated personal income/output ratio to produce the desired figure.

Prices are assumed to be constant, but if a price change is expected to result from the activity, then the input-output ratio has to be recomputed based on the new price before being used. Direct employment changes, changes in the average wage rate (perhaps due to a change in the size of the labor force), or other changes that are directly caused by the proposed activity should be examined. Any additional information indicating how the total wage bill changes with a change in an activity should be used, if possible. For example, business failures or disruptions caused by the activity and resulting in employment changes should be included. Other determinants of personal income, such as proprietor's income, dividends, interest, transfer payments, and other personal costs and revenues may be assumed to be changes in the same proportion as output revenue unless specific information indicates otherwise. Attempts should be made to assess these ratios whenever possible.

The change in disposable income equals the change in personal income minus the change in personal tax payments. Assuming a constant effective income tax rate (due to small incremental changes in income), this rate times the change in personal income gives the total income tax change. The tax rate may be obtained from the tables of "Effective Rates of State Personal Income Taxes for Selected Adjusted Gross Income Levels, Married Couple with Two Dependents, by State" or "Local Income Taxes, Rates, and Collections" in *State and Local Finances*. Changes in taxable property, together with the pertinent rate, give the property tax alteration.

The change in personal consumption is determined by a rough calculation of the coefficient of consumption applied to the change in disposable income. Thus, the proportion of disposable personal income spent on personal consumption expenditures, calculated from national data if local information is missing, may be assumed to apply to local disposable income. The *Statistical Abstract of the United States*, in the table on "Personal Income and Disposition of Income" gives the pertinent data from which the necessary coefficient can be calculated for the appropriate data (approximately 0.9 for all years). Multiplying this by the change in disposable incomes provides an estimate of the initial change in consumption.

The most difficult data requirements involve the identification of all activities linked to the proposed activity through an input-output relationship, and the determination of each coefficient indicating the dollar change in the supplying firm's (or individual's) output due to a unit change in the project activity (where this output relates to the particular activity being investigated). This information can come only from a detailed examination of project activity.

Evaluation and Interpretation of Data. The interpretation of these data must be based on exercised individual judgment. Judgments regarding high or low impacts should be made by persons performing the assessment. The reason for the judgment should be stated, also.

Special Conditions. The analysis can be improved if a complete input-output analysis is completed together with a detailed economic analysis of the change in personal income (and then in personal disposable income) that results from a change in the output of economic activities linked to the project. Where data are uncertain, an attempt should be made to use expected values, if possible.

Geographical and Temporal Limitations. The geographical area within which the change in consumption occurs must be determined a priori, but it should be defined by the spatial distribution of the affected labor force. Where project activities affect consumption outside the area (and hence, would not normally be included in the analysis), efforts should be made to separate locally important effects from the effects that are far removed from the project's impact.

The attribute measurement methodology presented assumes an average of the total annual changes over the lifetime of the project or activity. This is an arbitrary procedure, and temporal trade-offs (time discounting) can be applied if desired. Calculations can be made for different years in the future in terms of with or without project changes, and the separate figures are aggregated by first multiplying them by arbitrarily assigned normalized weights. Another simple alternative is to choose a single future year in which to compare with and without project changes, implicitly weighting all other years as zero.

Mitigation of Impact. Any detrimental impacts can be mitigated best if direct linkages are established with area industries, businesses, or other economic activities, encouraging an inflow of money into the local economy.

Secondary Effects. See description given under *Regional Economic Stability.*

REFERENCES

1. *State and Local Finances, Significant Features* (1967 to 1972), Advisory Commission on Intergovernmental Relations, Washington, D.C.

RESOURCES

Natural resources include the land, air, water, vegetation, animal, and mineral resources which constitute our natural environment, and provide the raw materials and spatial settings which are utilized in developing our familiar man-modified environment. These resources may be nonrenewable, such as metals and fuels, or renewable, such as water. Nonrenewable resources are of particular interest, since their consumption or utilization represents a commitment that is potentially irreversible or irretrievable.

Since fuel resources hold a position of extreme importance, they are treated as a separate attribute. Also, since many of the other natural resources are discussed through other attributes (ecology, air, water, and land), another attribute emphasizes the remaining nonfuel resources which are utilized in either a natural or transformed state for products and materials in the development of the human environment. A third attribute considers the aesthetic qualities of natural and man-modified environments—modified through the use of natural resources.

These attributes are summarized as follows:

Fuel resources
Nonfuel resources
Aesthetics

Fuel Resources

Definition of the Attribute. Fuel resources include all basic fuel supplies utilized for heating, electrical production, transportation, and other forms of energy requirements. These resources may take the form of fossil fuels (oil, coal, gas, etc.), radioactive materials used in nuclear power plants, or miscellaneous fuels such as wood, solid waste, or other combustible materials. Solar and hydroelectric energy resources or other energy sources in a current developmental state are not considered in this context.

Activities that Affect the Attribute. Since energy consumption relies almost entirely upon fuel resources, it is probable that any activity that consumes energy consumes fuel resources as well. Actions requiring consumption of energy can be categorized into (1) residential, (2) commercial, (3) industrial, or (4) transportation activities.

Residential activities include space heating, water heating, cooking, clothes drying, refrigeration, and air conditioning associated with the operation of housing facilities. Also included is the operation of energy-intensive appliances such as hair dryers, television sets, etc.

Commercial activities include space heating, water heating, cooking, refrigeration, air conditioning, feedstock, and other energy-consuming aspects of building or physical plant operation. Facilities which consume particularly significant amounts of energy include bakeries, laundries, and hospital services.

Industrial activities which require large amounts of fuel resources include power plants, boiler and heating plants, and cold storage and air conditioning plants. Other industrial operations that require process steam, electric drivers, electrolytic processes, direct heat, or feedstock may impact heavily upon fuel resources.

Transportation activities involving the movement of equipment, materials or personnel require the consumption of fuel resources. The mode of transportation may include aircraft, automobile, bus, truck, pipeline, or watercraft.

Source of Effects. Most presently utilized fuel resources are limited to the supplies of existing fossil and nonfossil fuels at or beneath the earth's surface. The demand for these fuels in the United States far outstrips the production rates of domestic supplies; hence, much of the fuel resources consumed daily in the United States comes from foreign sources. This places a dependence upon these foreign sources which bears heavily upon economic stability, and has obvious strategic implications. Furthermore, known reserves of certain fuels—particularly natural gas—are limited, to the extent that unless conservation measures are effected immediately, these supplies will be consumed within a matter of a few years.

Variables to be Measured. The most important variables to be considered in determining impacts on fuel resources are the rate of fuel consumption for the particular activity being considered and the useful energy output derived from the fuel being consumed. Various units may be utilized in describing consumption rates: miles per gallon, cubic feet per minute, and tons per day are commonly used in describing the consumption of gasoline, natural gas, and coal, respectively. Similarly, the energy output of various fuel-consuming and energy-consuming equipment and facilities may be described in many different units—horsepower, kilowatt-hours, and tons of cooling are a few examples.

A common unit of heat, the BTU, may be applied to most cases involving fuel or energy consumption. The BTU is the quantity of heat required to raise the temperature of one pound of water one Fahrenheit degree. In the evaluation of transportation systems, for example, alternatives may be compared on a BTU per passenger-mile or a BTU per ton-mile basis.

Other variables of concern include the availability (short-term and long-term) of fuel alternatives, cost factors involved, and transportation distribution and storage system features required for each alternative.

Data on the consumption of fuel resources may be applied to almost any environmental impact analysis, but the depth and degree to which data is required depends upon the nature of the project under consideration. For an analysis of existing facilities or operations, sufficient information should be available from existing records and reference sources. Where alternative fuels or transportation systems are under consideration, additional background information may be necessary to evaluate not only efficiencies, but cost-effectiveness and long-term reliability.

How Variables are Measured. Because of the complexities in the nature of the variables discussed above, most are measured by engineering, resource, and other professionals, although the results may be applied by most individuals with a technical background.

Once the heat contents of fuels are known, comparisons may be made on the basis of the heat content of each required to achieve a given performance. An energy ratio (ER) can be established as the tool for comparison. The ER is defined as the number of BTU of one fuel equivalent to one BTU of another fuel supplying the same amount of useful heat:

$$\text{ER} = \frac{\text{amount of Fuel No. 1 used} \times \text{heat content of Fuel No. 1}}{\text{amount of Fuel No. 2 used} \times \text{heat content of Fuel No. 2}}$$

Determination of energy ratios requires careful testing in laboratory or field comparisons and usually yields reliable results when conducted under impartial and competently supervised conditions. These ratios have been determined in various tests and are summarized in such publications as the *Gas Engineer's Handbook.*[1]

The consumption of fuels on an installation may be determined from procurement and operational records. Measurements may be made by using conventional meters, gauges, and other devices.

Evaluation and Interpretation of Data. The use of fuel resource data will more than likely be used in an environmental impact analysis for the benefit of planners and decision makers for either (1) evaluating the alternatives where either fuel consumption or fuel-consuming equipment or facilities are involved; or (2) determining the aspects of fuel and energy conservation that exist with regard to on-going actions. These aspects include the irreversable and irretrievable commitments of resources resulting from the action, the short-term/long-term tradeoffs, and the identification of areas for potential conservation and mitigation of unnecessary waste.

As previously discussed, the analysis should include aspects of efficiency, availability, cost of fuel and support facilities (transportation, distribution, storage, etc.), and projected changes in these values which might occur in the future. Secondary effects should also be considered, e.g., environmental effects due to fuel production (mining, refining, etc.) and impacts on air and water quality from combustion and related pollution control measures.

Special Conditions. If the activity results in significant additional demands or waste of fuels already in short supply, public controversy may be expected to follow. Natural gas supplies, presently limited or unavailable in some areas, should be considered with special emphasis. Electric consumption, in most cases, bears directly upon fuel resources, the effects of which should be included in the analysis.

Geographical and Temporal Limitations. Concern for fuel resources typically peaks during summer (when air conditioning loads are high) and winter (when heating loads are high). Thus, installations in northern climates would be expected to have the greatest concern for heating fuels, while southern installations would be more concerned with heavy cooling requirements in the summer, although exceptions to this general trend may occur due to localized demands or geographical or climatic effects. Proximity to natural supplies also may play an important role in fuel selection, since transportation may affect the availability and economic desirability of certain fuels.

Mitigation of Impact. Mitigation of impacts directly and indirectly attributable to fuel resources fall into two categories. The first pertains to mitigation by alternate fuel selection, and is based on a number of complex variables—availability, cost, environmental effects, and pollution control requirements, to name a few. Other factors to be considered in the selection are the short-term/long-term effects of a particular choice, and the irreversible and irretrievable commitment of resources associated with the selection.

The second category of mitigation is associated with the conservation of fuel resources, regardless of the type or types of fuel being consumed. Such measures can be applied to new construction in the form of additional insulation and design to incorporate energy conservation features related to color, orientation, shape, lighting, etc. The conservation of energy can be applied to existing facilities in the form of added insulation and programs to reduce loads on heating, cooling, and other utility consumption. Likewise, in the operation and maintenance of equipment, steps may be taken to further reduce fuel consumption by increasing efficiencies through proper equipment maintenance, reducing transportation requirements, and scheduling replacement of old equipment with newer, highly-efficient models.

Secondary Effects. Conversion of fossil and nuclear fuels into useful energy can lead to secondary effects on the biophysical and socioeconomic environment. Air emissions occur during extraction, processing, and combustion processes. Water quality may be affected by spills, acid mine drainage, and thermal discharges. Land use impacts include loss of habitat, land disturbance, erosion, and aesthetic blight. Solid waste problems resulting from mining and production activities include leachates, radioactive wastes, slags, and tailings.

Nonfuel Resources

Definition of the Attribute. This attribute considers the nonfuel resources which are utilized in either a natural or transformed state for products and materials in the development of the human environment. Various nonfuel products are manufactured from fuel resources, and are included in the definition. Specific examples include wood and wood products, metals, plastics, and nonmetallic minerals and materials.

Activities that Affect the Attribute. Few, if any, activities do not depend on natural resources in some way. Any activity that consumes materials and supplies, requires equipment and machinery, utilizes land, or produces waste products may have an effect on natural resources. Various materials— lumber, aggregates, cement, steel, asphalt, etc. are utilized in construction and repair activities. Operation of facilities depends on equipment that is manufactured requiring metallic and nonmetallic parts and components. Land use may deny access to minerals or other resources. Water disposal may result in loss of valuable resources—materials that could effectively be recycled, reclaimed, or reused.

Source of Effects. In order to develop and maintain our present life styles, many nonrenewable resources are being consumed at rates which indicate depletion of many critical materials in a matter of only a few decades; in some cases, only a few years. Furthermore, some of these materials in short supply are controlled by foreign powers, which results in even further complications, as dependency and strategic implications become important.

The following is a partial list of critical materials which presently are of major concern:[2]

aluminum	titanium
chromium	cobalt
platinum	mercury
iron ore	tungsten
nickel	lead
natural rubber	columbium
manganese	vanadium
zinc	fluorspar
tin	copper
silver	phosphate
wood (timber)	

In addition to materials in short supply and the problems associated with consumption of resources, another factor should be considered along with resource and resource conservation—energy. The manufacture and production of products and materials from raw resources consumes vast quantities of energy. The conservation of these materials not only reduces the depletion potential, but can result in energy savings, as well.

Variables to be Measured. For the impact assessment procedure, a study should be made that will (1) identify the activities or points of consumption of natural resources, (2) indicate the consumption rates, and (3) reveal the quantities and content of wastes resulting from those activities.

How Variables are Measured. Qualitative determinations relating specific activities and resource consumption may be made on the basis of first-hand knowledge of the activities and their mode of accomplishment, and a general knowledge of resources and resource management. Once these relationships are identified, the information may be utilized repeatedly, as it would remain valid until changes in the activity or its mode of accomplishment would occur.

Consumption rates are somewhat more difficult to quantify, and technical expertise may be required. Depending on the kind of activity and the type of resource, the rates may be reported in such various terms as pounds per year, tons per day, rounds per mission, etc. Based upon purchasing data and other records, input-output models may be constructed which depict the total effect of the resource utilization.

Content and quantities of waste products resulting from an activity may be determined from field studies during which actual waste samples are classified and analyzed, or may be estimated on the basis of the same input-output models discussed above or by simpler procedures (e.g., emission factors).

Evaluation and Interpretation of Data. After determination of points of resource consumption, quantities involved, and waste products produced, an evaluation may be made of the total impact by considering each of the resources being consumed in light of its individual status—abundance, importance, availability, economics, origin, energy to produce, recycle potential, etc. Life cycle thinking should be incorporated, i.e., looking at an activity with regard to the resource requirements for the life of a project from origin to completion.

Special Conditions. Special conditions may arise due to resource availability and price that can affect natural resources markets. Natural scarcities do exist for many resources, and these availabilities can be further jeopardized by embargoes or other supply interruptions (e.g., strikes). Although prices can actually assist in resource allocation in a free market supply and demand situation, efforts to artificially increase prices through such means as cartels, price-gouging, or cartel-like actions may occasionally place specific resources in a position of increased importance.

Geographical and Temporal Limitations. Specific geographic considerations include the origin of specific resources and the strategic implications associated with resource control. Also, transportation consumes fuel resources and should be considered in choosing alternatives (e.g., specification of a particular type of wood or building product that is unavailable locally). Seasonal aspects affect some resources (vegetation, mining, etc.) but most temporal limitations on resources are artificially produced.

Mitigation of Impact. Adverse impacts on natural resources and resource consumption can be minimized by economizing on resource requirements, development and use of substitutes, and recycling of scrap materials. These mitigations all can be considered as forms of conservation resulting in the use of less raw material per unit of output. Specific programs might include recycling of tires, glass, paper, metals, petroleum waste, construction and demolition debris, and general solid waste. These areas not only provide potential for conservation of materials, but some may be used for energy conversion, resulting in fuel conservation as well.

Secondary Effects. In addition to energy consumption, other environmental effects may be related to the consumption of resources. Activities associated with the extraction, transportation, and processing of materials to produce the finished products may impact on air, water, land, and ecology. Other social and economic factors may be affected as well.

Aesthetics

Definition of the Attribute. The aesthetic attribute may be used to describe impacts on the environment which are apprehended through the senses—sight, taste, smell, hearing, and touch. Although treated in part in other attributes (e.g., odors in *Air* and the overall category of *Noise*), tolerance levels based on aesthetic criteria are often somewhat different, in addition to the fact that aesthetics perceptions generally require the consideration of all the senses simultaneously. Visual perception is perhaps the most familiar of the areas, and the ensuing discussion will emphasize visual aesthetics and natural and man-modified landscapes.

Activities that Affect the Attribute. Generally, any activity that will alter the quality or distinguishable characteristic of the perceived environment can be considered as having an effect on aesthetics. Visual perception may be altered by activities involving construction, forestry and recreation management, transportation, water resource and land use planning, and other activities involving landscape and scenic vista modification. Other aesthetic perceptions (hearing, smell, etc.) may be affected by industrial activities, burning, aircraft operations, waste discharges, and various facility operation and maintenance activities.

Source of Effects.   The activities that affect aesthetics do so by creating changes in the aesthetic characteristics of the environment as they are perceived by individuals (examples of characteristics include color, texture, scale, harmony, etc.).   These perceptions are explained more fully in the following sections.

Variables to be Measured.   Individual perceptions and values for defining beauty make it difficult to quantify aesthetic impacts.   Perception of ugliness, however, is more nearly agreed upon. In most cases, aesthetic criteria can be formulated by persons who have had experience in design and have acquired a sensitivity to the characteristics of the natural setting and man-made objects that make them pleasing or displeasing to the human senses.   Measurement techniques for identifying and describing aesthetic impacts are basically of two types:

1. Subjective: The qualitative analysis procedures based on the developer's best knowledge of design characteristics.
2. Objective: The quantitative analysis procedures based on established thresholds.   The essence of this methodology includes design standards, architectural controls, sign ordinances and landscape criteria.   As an example, natural landscape aesthetics may be analyzed using the variables as follows:

- *Landscape character* in terms of the landscape setting:
  a. Boundary definition: physical, vegetative, topographic, etc.
  b. General form and terrain pattern.
  c. Vegetational patterns.
  d. Features: hills, valleys, cliffs, promontories.
  e. Water and land interfaces: conditions and quality.
  f. Weather patterns.
  g. Cultural interfaces: artificial objects, transportation facilities, structures, etc.
  h. Natural and man-made acoustical features: sound absorption, falling water, birds.

- *Macro (major) components*[3]:
  a. Unity: the cohesion of the parts into a single harmonious unit, described by the presence or absence of a single dominant factor and complimenting subordinate elements, contributing to a pleasant total composition.
  b. Variety: diversity without confusion, more than one element contributing to richness; the maximum opportunity for visual stimulus.
  c. Vividness: quality lending to sharp visual impression—distinction.

- *Micro (minor) descriptive elements:*
  a. Texture: identifying quality or disposition of the vista (e.g., rocks, trees, grass, cultivated crop patterns, etc.), soft, sharp, flowing, rough.
  b. Color: may be described in terms of hue, lightness, and saturation.
  c. Contrast: diversity of adjacent parts in color, shape, or texture.
  d. Uniformity: similarity between features.
  e. Scale: proportion of one object compared to another, particularly important in considering man-modified landscapes.

- *Changing qualities:*
  a. Distance: proximity to components in the vista.
  b. Observer position: aesthetic qualities of a given area may vary with viewer location.
  c. Speed of observation: duration of viewer's observance.
  d. Time: daily and seasonal changes.
  e. Observer's state of mind: expectations, values, mood.

How Variables are Measured; Evaluation and Interpretation of Data.   Due to the nature of aesthetics and human perception, significant features are often difficult to quantify. Many methods, however, have been developed in an attempt to establish standards of comparison, such as to arrive at a basis for determining which type of landscape (for most persons) is desirable or more desirable than another. These methods take two general forms:

1. A relative numerical weighting of each of the various intrinsic and extrinsic landscape resources, as individual components and as a composition reflecting presence and relationships of the descriptive elements listed above. These procedures attempt at quantifying visual relationships, place a value on aesthetic resources, and describe the implications of changes on the landscape in terms of scenic quality, as ranked with other environmental changes.
2. The non-numerical methodologies tend to place emphasis on ranking of visual attributes according to the same elements as the numerical scheme, but evaluate the aesthetic elements in terms of comparative analysis

based on established criteria. They do not assign numerical weights, but may, in some cases, assign a position on negative value.
In addition, most studies can be categorized as:

- The *visual analysis methodologies:* visual components of environment are inventoried and assessed by the planning staff or decision makers.
- The *user-analysis methodologies:* designed for attempting to find out how the general public feels about various aesthetic and potential impacts. Used as inputs to above assessment.

Special Conditions. Since the value, importance, or expression of beauty is relative to the variable of perception, it is important to note that the following conditions bear significantly on the degree of aesthetic impact:

- The observer's state of mind: factors of current perceptual setting and environmental life style, coupled with past experiences and future expectations, can produce varying impressions of aesthetic quality.
- The observer's background: cultural, economic and social background can determine perceived aesthetic qualities.
- Context of the observation: the setting of an aesthetic observation may bear upon its acceptability, e.g., is it "out-of-place"?

Several sources of information are available for further study and evaluation of aesthetic related impacts.[3,4,5]

Mitigation of Impact. Aesthetic impacts are frequently of a controversial nature. While it is generally agreed that everyone would like to enjoy clean air, pristine waters, scenic vistas, and serenity in their everyday living, economics and other "facts of life" do not always make this possible. However, most adverse aesthetic impacts may be minimized once an aesthetic inventory is provided to planners and designers, so that desirable features associated with a project might be maintained and enhanced or incorporated into the project, and undesirable features of the project redesigned.

Secondary Effects. Aesthetic qualities may be associated closely with land use characteristics—an association leading to potential secondary impacts on almost any other biophysical or socioeconomic attribute. Aesthetic impacts not only reflect upon psychological needs, but frequently may be related to land prices, economic security, and community needs.

REFERENCES

1. *Gas Engineer's Handbook*, Chapter 22, "Fuel Comparisons," The Industrial Press, New York, 1965.

2. "Critical Imported Materials," Council on International Economic Policy, USGPO, Washington, D.C., December 1974.

3. "An Aesthetic Overview of the Role of Water in the Landscape," by R. Burton Litton *et al.*, prepared for the National Water Commission by the Department of Landscape Architecture, University of California, Berkeley, July 1971.

4. "Aesthetics for Environmental Planning," by M. D. Bagley *et al.*, for the Environmental Protection Agency, Report No. EPA - 600/5 - 73-009, November 1973.

5. *Environmental Psychology—Man and His Physical Setting*, Harold M. Proshansky *et al.* (eds.), Holt, Rinehart and Winston, New York, 1973.

Appendix C

A STEP-BY-STEP PROCEDURE FOR DEVELOPING EIA/EIS

There currently exist many methodologies for impact assessment, as evidenced in Chapter 6. This appendix details a systematic approach to the development of an environmental impact analysis, utilizing a basic matrix approach and the multidisciplinary attribute descriptor package from Appendix B. Although the matrix method is employed, similar procedures utilizing the attributes and basic concept could be developed.

The following steps, as shown in Fig. C-1, detail the procedure to be used in the preparation of an environmental impact analysis. The degree of consideration to be exercised within some of the steps may vary with project scope and magnitude, but the basic algorithm is applicable in all cases.

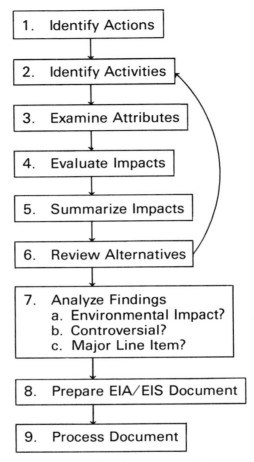

Fig. C-1 Step-by-step procedure.

Step 1. Identify Actions that Require Analysis.

All federal agencies, and many state and local agencies, are required to consider the environmental effects of implementing their major programs. Therefore, it is first necessary to decide whether the action is major, or one that may produce significant effects on the environment. CEQ guidelines require federal agencies to develop specific criteria for identifying those actions *likely* to require environmental statements and those actions *likely not* to require environmental statements. These normally take the form of lists, and are publicly available.

The first step is to compare the project under consideration with those on the list of projects likely to have a significant environmental impact. If the project appears on the list, the user should proceed with Steps 2 through 6, in detail. If it does not, and there are no obvious severe impacts, the user should proceed with Steps 2 through 6, in lesser detail, to verify the absence of significant impacts.

As an example, assume that scheduled programs at an Army installation include construction of 200 units of family housing. Examination of Fig. C-2 indicates that this action corresponds to item No. 20 in the list of representative Army actions that might be assessed to determine whether or not it will be environmentally significant or controversial.

Step 2. Identify Relevant Project Activities

Identify detailed activities associated with implementing the project or the program: agency activities may be categorized into functional areas. For each functional area, detailed activities associated with implementing projects or programs may be developed. The user should supplement these activities with project-specific activities.

In the case of the construction of 200 family housing units, the construction activities shown in Fig. C-3 may be utilized. Those activities not applicable to the project should be crossed off, and supplemental activities should be added to encompass the project-specific requirements.

Step 3. Examine Attributes to be Reviewed

The user should examine the attributes in Appendix B and become familiar with the general nature of the individual attributes and the kinds of activities that may impact on them. In addition, the descriptor packages may be used to identify areas where available technical expertise is deficient and additional assistance may be required.

Step 4. Evaluate Impacts Using Descriptor Package and Worksheets

Using the activities (as developed in Step 2) and the attribute list, construct a matrix worksheet, with activities on one axis and environmental attributes on the other. Figure C-4 indicates an example format. Use the attribute descriptor package (Appendix B) to:

• Identify potential impacts on the environment by placing an "X" at the appropriate element of the worksheet.

1. Development or purchase of new type of aircraft or other mobile facilities or substantially modified propulsion system.
2. Development or purchase of new weapon system.
3. Real estate acquisition or outleases or permitting or exchange or disposal of real estate.
4. Major construction projects.
5. New installations.
6. Production, storage, relocation, or disposal of chemical munitions, pesticides, herbicides, and containers.
7. Use of pesticides or herbicides, when proposed for use other than in accordance with the label as registered.
8. Harvesting of timber, wildlife, etc. (significant amounts).
9. Intentional disposal of any substances in a significant quantity or on a continuing or periodic basis.
10. Mission changes and troop deployments which increase or decrease population in any area.
11. Major research and development projects and test programs associated with R&D projects.
12. Any action which, because of real, potential, or purported adverse environmental consequences, is a highly controversial subject among people who will be affected by the action, or which, although not the subject of controversy, is likely to become a highly controversial subject when it becomes known to the public.
13. New, revised, or established regulations, directives, or policy guidance concerning activities that could have an environmental effect (e.g., training, construction, or mission change). Regulations, directives, or policy guidance limit any of the alternative means of performing the actions on this list. Broad programs which could indirectly affect other actions on this list (e.g., Volunteer Army Program, Energy Conservation Program, or Expansion of the Women's Army Corps).
14. Intentional disposal of any materials in the oceans or other bodies of water.
15. Large quarrying, timbering, or earth-moving operations.
16. Airfield and range operations for test or training purposes.
17. Constructing, installing, or maintaining fences or other barriers that might prevent migration or free movement of wildlife.
18. Approval of new sanitary landfills, incinerators, and sewage treatment plants and operation of existing facilities.
19. Existing or changes to master plans.
20. Construction or acquisition of new family housing over 25 units.
21. Dredging.
22. Exercises involving divisional or larger units on or off Federal property, or where significant environmental damage may occur regardless of unit sizes.
23. Exercises involving smaller units when the training involves non-Army property or there is a significant amount of heavy or noisy equipment involved.
24. New deployment or relocation or disposal of nuclear power plants.
25. Operation of existing or new government-owned production facilities.
26. Ammunition storage facilities, new or continuing operations, or transportation of ammunition.
27. Closing or limiting of areas that previously were open to public use; that is, roads or recreational areas, etc.
28. Activities that will or may increase air or water pollution or disrupt plant life on the real estate.
29. Construction on flood plains or construction that may cause increased flooding.
30. Fuel conversion or continued consumption of significant quantities of fuel in short supply.
31. Increase in energy requirements.
32. Channelization of streams.
33. New facilities for aircraft, increase in number of aircraft at existing fields, and operation of existing aircraft in significant numbers.
34. Activities in wetland areas.
35. Storage, use, and disposition of POL products.
36. Use, storage, and disposition of radioactive materials, other than as authorized in Title 10, code of Federal regulations.
37. Operation and maintenance of power-generating equipment.
38. Control of pest organisms such as birds or other animals.
39. Construction of roads, transmission lines, or pipelines.
40. Award or termination of major contracts for supplies of natural resources; e.g., coal, oil, etc.
41. Transportation and testing of chemical agents and munitions.
42. Determination of safety standards, especially quantity-safety distances.
43. Development or purchase of new types of equipment, other than mobile facilities and weapon systems.
44. Outdoor large-scale or controversial testing of newly developed systems or material.
45. Continued operation of existing facilities which are causing pollution.

Fig. C-2   Representative Army actions that might have a significant environmental impact.

Site Access/Delivery
  Railroad
  Road
  Water
  Air
  Pipeline

Support Facilities Operation
  Asphalt plant
  Aggregate production
  Concrete operations
  Foundry & metal shop
  Fuel storage and dispensing
  Material storage
  Personnel support
  Utilities provision
  Solid waste disposal
  Sewage disposal

Site Preparation
  Clearing and grubbing
  Tree removal
  Existing structure removal
  Demolition debris disposal

Excavation
  Topsoil stripping
  Excavation
  Backfill
  Channeling and Dredging
  Hauling

Quarrying & Subsurface Excavation
  Cutting and drilling
  Loosening
  Hauling
  Drainage

Foundations (Buildings and Roads)
  Base course
  Footings
  Compaction
  Piling
  Foundation Mats
  Groundwater control

Bituminous Construction
  Hauling
  Mixing
  Placing and spreading
  Compaction
  Curing and sealing

Concrete Construction
  Hauling
  Mixing
  Placing
  Finishing

Masonry Construction
  Hauling
  Forming
  Mortar mixing
  Placing
  Finishing

Steel Construction
  Hauling
  Erecting
  Finishing

Timber Construction
  Hauling
  Pest/Insect protection
  Cutting and shaping
  Erecting
  Finishing

Finishing — General
  HVAC (heating, ventilating
    and air conditioning)
  Electrical
  Plumbing
  Cleanup operations
  Landscaping
  Painting

Fig. C-3   Basic construction activities.

- Collect baseline data on the impacted attributes.
- Quantify the impact where possible.

For instance, construction of the 200 unit family housing project might require large scale excavation which might cause erosion, increased suspended solids in the receiving waters of the stream and decreased dissolved

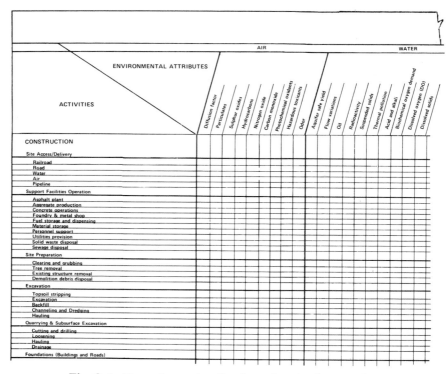

Fig. C-4   Example construction impact evaluation matrix.

oxygen. Mark an "X" on the worksheet for all such negative potential impacts and a "+" for any positive potential impacts. It should be emphasized again that this evaluation is to be done on an interdisciplinary basis.

Step 5. Summarize Impacts

For potential impacts, mark with "X" or "+" on the worksheet; summarize the impacts using Fig. C-5. Shade the areas of net positive or net negative impacts, using the shading intensity to indicate the significance of impact. For example, for impacts on erosion, suspended solids, and dissolved oxygen, evaluate the magnitude of the project and the site characteristics, use the scientific information provided in Appendix B, and determine the degree of severity of the impact on each attribute. Finally, summarize the impacts on each attribute from all project activities by using the key shown in Fig. C-5.

Category	Attribute Number	Attribute
Resources	49	Aesthetics
	48	Nonfuel Resources
	47	Fuel Resources
Socioeconomic — Economic	46	Per capita consumption
	45	Public sector revenue
	44	Regional economic stability
	43	Community needs
Socioeconomic — Human	42	Physiological systems
	41	Psychological needs
	40	Life styles
	39	Social behavior effects
Sound	38	Performance effects
	37	Communication effects
	36	Psychological effects
	35	Physiological effects
Ecology	34	Aquatic plants
	33	Natural land vegetation
	32	Threatened species
	31	Field crops
	30	Fish, shell fish, and water fowl
	29	Small game
	28	Predatory birds
	27	Large animals (wild and domestic)
Land	26	Land use patterns
	25	Natural hazard
	24	Erosion
Water	23	Fecal coliform
	22	Aquatic life
	21	Toxic compounds
	20	Nutrients
	19	Dissolved solids
	18	Dissolved oxygen (DO)
	17	Biochemical oxygen demand
	16	Acid and alkali
	15	Thermal pollution
	14	Suspended solids
	13	Radioactivity
	12	Oil
	11	Flow variations
	10	Aquifer safe yield
Air	9	Odor
	8	Hazardous toxicants
	7	Photochemical oxidants
	6	Carbon monoxide
	5	Nitrogen oxide
	4	Hydrocarbons
	3	Sulphur oxides
	2	Particulates
	1	Diffusion factor

*Net Positive Impact +

ATTRIBUTE NUMBER

Net Negative Impacts X

□ No Significant Impact

▨ Moderate Impact

■ Significant Impact

Project Name _____

Project Number _____

Alternative _____

*Positive impacts are shown above the attribute number and negative impacts below.

Fig. C-5 Summary of Impacts.

Step 6. Review Other Alternatives

Repeat the procedure for the alternatives considered.
Examples of alternatives for the family housing project example may be:

- No action alternative
- Alternatives related to different designs and/or projects and activity site
- Alternative measures to provide for compensation of fish and wildlife losses

Step 7. Analyze Findings

Determine the answers to the following questions concerning the action:

- Will the implementation of the program have a significant adverse effect on the quality of the environment?
- Will the action be deemed environmentally controversial?
- Is the action a major line item in the budget? (for federal actions)

If the answer to any of these questions is "yes," the proponent is required to prepare an EIS; otherwise, an EIA is normally filed. If it is decided that an EIS is not required for a project or activity of a type described in the published agency list of projects normally requiring an EIS, a "negative declaration" (ND) must be prepared for public record setting forth this decision and the reasons for the determination.

While the impact analysis worksheets, summary sheets, and attribute descriptor package assist in pinpointing environmental impacts of a scientific or technological nature, determining the potential for public controversy requires a more subjective approach. (Refer to *Special Issues* and *Public Participation* in Chapter 8.)

Step 8. Prepare Analysis Document

Depending on whether the decision is to prepare an EIA, an EIS, or an ND, the analysis should be documented as indicated in Chapter 3.

Step 9. Process Documented Analysis

Depending on whether the document is an EIA, EIS, or ND, specific directives for processing or retaining the document are issued by each agency. These directives should be consulted for guidance in processing and coordinating the analysis document.

Index

Index

Acid, 216–217
Ad hoc methodologies, 73
Adverse impacts, 30–31
Aesthetics, 60–61, 312–16
Agency activities, 37–38
Aggregation, 67–68
Agricultural Stabilization and Conservation Service, 259
Air, 38–44, 170–201
  carbon monoxide, 187–89
  diffusion factor, 170–73
  hazardous toxicants, 192–95
  hydrocarbons, 180–184
  nitrogen oxides, 184–87
  odors and, 195–200
  particulates and, 173–76
  photochemical oxidants, 189–92
  sulfur oxides, 176–80
Air pollution, 40–44, 170–201
Air pollution control, 154
Algae, 221
Alkali, 216–17

Ambient standards, 40
Animals, large, 245–48
Aquaculture, 216
Aquatic life, 225–26
  water and, 47
Aquatic plants, 265–67
Aquifer safe yield, 203–208
ASCS. *See* Agricultural Stabilization and Conservation Service
Assessment, 18–19
Atmosphere, 48
Atomic Energy Commission, 4
Attribute descriptor package, 168–315
Ayres, R. V., 137

Bald eagle, 250, 261
Bioassay, 224, 225
Biochemical oxygen demand, 217–18
Biosphere, 48
Black footed ferret, 261
Blue baby disease, 220
BOD. *See* Biochemical oxygen demand
BTU, 146

Date Due

Due	Returned	Due	Returned
FEB 14 '78	FEB 14 '78		
MAR 3 '78	MAR 5 '78		
JUN 3 '78	MAY 19 '78		
JUN 2 '78	JUN 5 '78		
DEC 8 '78	DEC 7 '78		
JUN 9 '79	APR 23 '79		
JUN 1 '79	JUN 21 '79		
NOV 28 '79	NOV 26 '79		
APR 30 1983	MAY 06 1983		